SHERLOCK HOLMES

AND

A QUESTION OF SCIENCE

By

Christopher F. Lindsey

HADLEY PAGER INFO

First Edition 2006

ISBN 1-872739-16-4
(978-1-872739-16-8)

Copyright © Christopher F. Lindsey   2006 (Text)

All rights reserved, including translation. This publication, or any part of it, may not be reproduced, stored in a retrieval system or transmitted in any form without the prior written permission of the publisher.
The cover design is the copyright of Hadley Pager Info and licensors

Printed and bound in England by Cromwell Press Ltd.,
Trowbridge, Wiltshire

HADLEY PAGER INFO
Leatherhead, Surrey, England

FOREWORD

As a dedicated Sherlock Holmes enthusiast myself, I have greatly enjoyed researching the content for this book, which concentrates on the scientific and technological aspects of the adventures about Holmes and his faithful medical companion Doctor Watson. I have included many quotations to illustrate the surprisingly wide range of scientific topics appearing in the stories. Numerous references have also been given throughout the text to direct the reader to the comprehensive notes at the end of each chapter, which I hope will both inform and elucidate. Inevitably key details of some of the cases have been revealed in several of the quotations or accompanying comments appearing within the text, and as I would not wish to spoil anyone's enjoyment, readers are recommended to read the stories first in order to obtain a full understanding of the accomplishments of the famous detective and his colleague.

Many of the intriguing crimes that Holmes and Watson had to investigate either occurred in London or originated in the metropolis, the trail often starting at their rooms in Baker Street. With this strong London background to the adventures, it is obvious that enthusiasts who are able to visit the capital will wish to see locations associated with the detective and the Doctor. As the city has changed a lot since the Holmesian period of the late 19th and early 20th centuries, it may be more rewarding for devotees to visit places where the detective's memory is kept alive. At the top of this list would probably be the Sherlock Holmes Museum in Baker Street, with room scenes from the stories, and specialist shop selling a range of suitable memorabilia. A follow up, especially for those requiring a drink or a meal, would be to search out The Sherlock Holmes public house (with appropriate sign outside) in Northumberland Street, not far from Trafalgar Square. Here can be seen an interesting collection of Holmes memorabilia in the bar, and a version of the famous study at 221B Baker Street, which has been re-created upstairs in the restaurant.

Another place to visit is Smithfield, where the enthusiast can view the outside of St. Bartholomew's Hospital – the location where Holmes and Watson first met before deciding to share rooms together. Here there is a museum (open to the public), which is devoted to the hospital's long history since its foundation in 1123.

Other places which should be mentioned and have a specific scientific interest are the Natural History Museum in South Kensington and the London Zoo in Regents Park. These two institutions house many, if not all, of the creatures mentioned in the adventures, but as their collections are large, there may be quite a search to find them. At the same time the Science Museum in Exhibition Road will also have some appeal, as old scientific machines, apparatus, trains and road vehicles used in the Victorian era can be found there.

As related in several of the adventures, Holmes and Watson often commenced their journeys out of London from the mainline stations, and although these have altered greatly since the days of steam trains, they can still be visited. A ride can also be taken on the Metropolitan Line to Aldgate, a route that plays such an important role in The Bruce-Partington Plans. If you start your journey at Baker Street Station, you will be able to see wall tiles bearing the familiar silhouette of the famous detective inside the station, while outside the main entrance in Marylebone Road, is a fine tall modern statue of Holmes, complete with pipe and deerstalker.

Christopher F. Lindsey
London

ACKNOWLEDGEMENTS AND NOTES FOR READERS

I am greatly indebted to all those who have helped me during the preparation of this book. Alan Lindsey has provided valuable editorial support, and has given assistance in the selection of illustrations for the various chapters. Valerie Naggs and Myrtle Bloomfield have prepared some of the illustrations included. Myrtle Bloomfield has also helped me in part of the research work, as has Pauline McLaren and Hazel Lindsey. To all of them I express my profound thanks.

It has not been possible to list individually the diverse sources consulted in the compilation of this book, but here I would like to acknowledge the helpful assistance of librarians, and numerous authors and compilers whose works have provided information for some of the chapters and many of the notes in my book.

The Sherlock Holmes stories ran out of copyright in the United Kingdom and in the European Union in 2000, although some of the later stories remain in copyright within the USA. The Conan Doyle Estate has Trademark registered many of the characters appearing in the Sherlock Holmes stories and the existence of these trademarks is hereby recognized.

It should be noted that occasionally some quotations from the original stories have been shortened or combined, have had words omitted or inserted or have been otherwise slightly modified, with the aim of providing better grammatical readability of the text. Some spellings, punctuation and hyphenation may also differ slightly from the original published editions.

Christopher F. Lindsey

ABOUT THIS BOOK

Christopher Lindsey has written a number of biographical accounts of British scientists, engineers, and doctors of medicine, which have been published in the *Oxford Dictionary of National Biography*. The Publishers of this book were therefore very pleased that when presented with the suggestion that he should examine Sherlock Holmes as a scientist he took up the idea with enthusiasm. This has resulted in a book to delight all those who enjoy the Sherlock Holmes stories. It provides a well researched and well referenced account of the wide range of scientific knowledge and observation that Sherlock Holmes deployed in solving the various murders and mysteries so ingeniously developed by his creator Sir Arthur Conan Doyle. The book also provides a picture of the earliest developments in forensic science in the late Victorian period, and combined with its valuable collection of notes fulfils a long standing need for a source book on all the scientific aspects of the Sherlock Holmes stories.

ARTHUR CONAN DOYLE (1892)

CONTENTS

	Page
Foreword	7
Acknowledgements And Notes For Readers	9
About This Book	10

1. **INTRODUCTION** — 13

2. **THE MEDICAL RECORD** — 28
   (Medicine and Pharmacology)

3. **THE OLD, THE PRECIOUS AND THE BURIED** — 50
   (Archaeology, Geology and Anthropology)

4. **LADY NICOTINE AND THE THREE PIPE PROBLEM** — 65
   (Tobacco and Smoking)

5. **BULLETS, SHARP BLADES AND BLUNT INSTRUMENTS** — 80
   (Ballistics and Weaponry)

6. **THE ACID TEST** (Chemistry) — 96

7. **THE CRYPTIC MAZE** — 112
   (Calligraphy and Cryptography)

8. **THE UNIVERSE AND NUMBERS** — 124
   (Astronomy, Meteorology and Mathematics)

9. **SEARCH FOR THE TRUTH** — 132
   (Optics and Photography)

10. **TECHNOLOGY PLAYS ITS PART** — 140
    (All Things Technical)

11.	ANIMAL OR VEGETABLE ? (Natural History)	161
12.	"YOU KNOW MY METHODS WATSON !" (Deduction and Forensic Science)	187
	List of Sherlock Holmes Stories	201
	Illustrations: Source and Acknowledgements	203

1 INTRODUCTION

Sherlock Holmes is without doubt the world's most famous and ingenious detective to appear within the pages of crime fiction. Sir Arthur Conan Doyle, who was the originator and author of the tales about the great detective, based his sleuth on Doctor Joseph Bell, his former tutor at the medical school of Edinburgh University, where Doyle himself once trained to become a doctor. Bell, both as surgeon and lecturer, greatly impressed his students and others by his deductive powers based on close observation and the examination of all known facts about a patient. It is said that he could frequently diagnose the patient's illness before being informed of the problem. Conan Doyle decided to give his eminent detective some of the distinctive attributes of Doctor Bell. To these were added certain embellishments and eccentricities which would create a character both complex and unusual.

From the numerous descriptions of him which appear in the adventures, readers can really get to know Sherlock Holmes, so that he comes to life in their imaginations. For example we are told that he is tall and thin with a prominent hawk-like nose, has grey eyes, speaks in a high voice and has a sharp, highly intelligent mind. He is also strong, athletic, and an excellent boxer. His interests, apart from his dedication to criminal investigation, include science, writing, and the appreciation of music. He is also an accomplished violin player. An enthusiast of all forms of smoking (a habit common in Victorian and Edwardian Britain) – Holmes maintains that it aids his thought processes when on a case; but his main vice is drug taking, an addiction that was frowned on then as it is today. His dress usually consists of a frock coat or of tweeds, and for outdoor use he sometimes wears an ulster [1A]. His cap with ear-flaps (never named as a deerstalker [1A]) he wears on visits to the country.

As a confirmed bachelor, his attitude to women is both complicated and variable, and it seems he holds a whole gamut of views about them, ranging from considerable mistrust to deep respect. Irene Adler above all, is held in the greatest esteem, not just because she gets the better of him, but also for her good looks. On one occasion he describes her as *the daintiest thing under a bonnet on this planet"* (A Scandal in

Bohemia). High praise indeed.

Holmes has a brother called Mycroft, who appears in a few of the cases (e.g. The Greek Interpreter; The Bruce-Partington Plans). He is also highly intelligent and demonstrates abilities which are similar and possibly superior to those of Sherlock. In one adventure (The Final Problem) he even takes on the role of reliable coach driver for Doctor Watson.

Further descriptive details about Holmes, who, among his many accomplishments, is also a master of disguise, can be found by reading all the stories and novels about him.

Mycroft

Having decided upon a suitable detective to carry out the investigations, it was necessary to give him an assistant with different skills to complement his work. What better than to introduce a man with medical qualifications? Accordingly Doctor John Watson, M.D., was created by Conan Doyle as the affable companion and recorder of the completed cases undertaken by Sherlock Holmes. With Watson on hand to provide all the medical expertise required in investigations littered with murders, attempted murders and injuries and diseases of all types, Holmes was to have the ideal partner. Also, as Conan Doyle himself was a doctor, we can easily believe that the medical element, which is so important in many of the adventures, has been authentically portrayed.

Perhaps at this stage, we should pause and consider an important question. What is it exactly that readers find so fascinating about Sherlock Holmes and his cases? Inevitably there are many answers to this question.

THE WRITING AND SETTING

First of all the quality of the writing cannot be disputed. It invariably holds the reader's attention throughout, and although there are occasional verbose passages which could have been shortened, the reader is still encouraged to go on turning the pages to find out what happens. Of course many of the stories were first published in magazines such as *The Strand* [1A] which regularly attracted purchasers by the inclusion of this type of exciting adventure.

In addition, the Victorian and Edwardian settings of the stories are intriguing, with their echo of a past age. Many of the events that are recorded take place in a London with stations served by steam trains, and roads crowded with trundling horse-drawn hansom cabs [1A], growlers [1A] and other vehicles. In some of the mysteries (e.g. The Copper Beeches; The Sign of Four; The Bruce-Partington Plans), there is an occasional swirling winter fog to add to the atmosphere. Other snippets from the past inform us of everyday life. For example we learn of the high level of service provided to householders by the postal authorities, with frequent deliveries throughout the day. There are also many editions of the London evening newspapers with their useful 'classified advertisement' columns, which Holmes uses at times to further an investigation.

THE PLOTS

The plots of the stories are also ingenious and some are quite strange: The Red-Headed League is a good example. Most are fascinating and baffling to the reader until Holmes explains the solution. When Conan Doyle wrote a story, he had to devise the ending before filling in the detail which would lead to a satisfactory conclusion; although for Holmes, the case does not always end in success. He actually fails in a few of the mysteries [1A], and this adds to the realism of the writing by demonstrating that a perfect result, just as in real life, is not always attainable.

THE CHARACTERS AND LOCATIONS

A whole range of people of all classes, professions and backgrounds appear in the adventures, with some of them visiting Holmes in his consulting room at 221B Baker Street. This room, which he shares much of the time with Doctor Watson, is untidy and unusual in that it incorporates a laboratory for his chemistry experiments, and a record filing system which is not always up-to-date and must sometimes have required serendipity to retrieve the appropriate information. The room also contains some surprises. Cigars are kept in the coal-scuttle [1B], tobacco in the toe of a Persian slipper [1B], and the wall has bullet-pocks from occasional target practice.

221B Baker Street

HOLMES' RECORD KEEPING

In The Musgrave Ritual we learn something about Holmes' erratic approach to his own record keeping and the disorganized way in which he keeps his chemicals and criminal relics. Watson writes that they *had a way of wandering into unlikely positions, and of turning up in the butter-dish, or in even less desirable places.* Holmes *had a horror of destroying documents, especially those which were connected with his past cases.*

These untidy records are a bugbear to Watson, who goes on to explain that Holmes would only sort them out once every year or two. Watson's frustration is understandable because the *papers accumulated until every corner of the room was stacked with bundles of manuscripts.*

THE HOUSEKEEPER

Holmes and Watson are looked after much of the time by the efficient Mrs. Hudson, their housekeeper and landlady, who has to contend with the comings and goings of many people, as well as the peculiar ways of her two lodgers, and of Holmes in particular. The Dying Detective explains the effect Holmes has on Mrs. Hudson, and mentions as well *his weird and often malodorous scientific experiments.* Also Mrs. Hudson is upset by the visits of the young Baker Street Irregulars [1C], Holmes' unofficial but highly effective ragamuffin 'police force'. The Baker Street Irregulars are an inventive touch which add to the appeal of the adventures.

THE POLICE

Holmes' relationship and dealings with detectives from both the Metropolitan Police and other forces is usually congenial, and although he outshines them both in capability and success, he is very generous in allowing them to take the credit when a case is satisfactorily solved. However members of the police vary in their attitude to Holmes and some are quite critical of his 'methods'.

THE CHIEF VILLAIN

Every detective novel or story has to have one or more perpetrators of the murders and crimes. The Sherlock Holmes stories have these in good measure but with the added ingredient in some of the stories [1B] of an arch-criminal named Professor Moriarty [1B]. Whenever he plays a leading role in an adventure, or is just a shadowy 'Mr. Big' in the background, he is deemed to be brilliant (on a par with Holmes), and is the declared evil opponent of the great detective. If Conan Doyle had continued writing these adventures, he may well have expanded the role for Moriarty. A tussle of minds

Professor Moriarty

will always provide an entertaining and gripping plot for detective stories of this type.

SHERLOCK HOLMES' SKILLS AT DEDUCTION

The story-lines have diverse methods for the murders or attempted murders of the victims, and Holmes has to employ his deductive skills to the full to carry out the investigations. In time he becomes famous for the 'scientific methods of deduction' that he uses to solve them.

FIRST METHOD OF DEDUCTION

These scientific methods can really be divided into two types. The first one is logical deduction by using the facts and evidence available, in other words the processing of data. This includes the systematic search for clues, often with Holmes utilizing his famous magnifying lens; the conducting of interviews with witnesses and injured victims; and also sometimes examining information about the case supplied by the official police. *"One forms provisional theories and waits for time or fuller knowledge to explode them,"* says Holmes to Mr. Ferguson in The Sussex Vampire, and then goes on to add that it is a bad habit, *"but human nature is weak"*.

The opening chapter of The Sign of Four goes into some detail about the investigative techniques of Holmes. In it he admirably demonstrates the first method of deduction by studying a watch which Watson hands over to him. The latter is both surprised and upset by Holmes' accurate description of the watch's former owner, who turns out to be Watson's deceased brother. In protest, he declares: *"You have made inquiries into the history of my unhappy brother, and you now pretend to deduce this knowledge in some fanciful way. You cannot expect me to believe that you have read all this from his old watch!"*

Scientific methods are also mentioned in Black Peter. This case is brought to Holmes by Stanley Hopkins, a young police inspector who has the respect of a pupil for the scientific methods of the famous amateur. It is odd though that Conan Doyle calls Holmes an 'amateur'. Considering the detective is successful in numerous cases, and that he seems to obtain an income from detection, Holmes is surely an accomplished professional, albeit a private one as compared to one in the established police force. Later in the same story Hopkins tells Holmes that on finding the body of Peter Carey: *"I know your methods, sir, and I applied them."* Hopkins then proceeds to tell him that there were no footprints, and Holmes queries this fact: *"Meaning that you saw none?"* Holmes then points out that there are invariably some clues left at a crime scene *"which can be detected by the scientific searcher"*.

In so many of the cases he pursues, Holmes emphasizes the importance of investigation and analysis when considering a fact or material clue in order to arrive at a correct result. To aid progress towards the solving of a mystery, throughout the adventures Holmes is always stressing the value of data, *"ah, I have no data"* (The Copper Beeches) and again in the same story: *"data, data, data"*. And *"I have no data yet"* (A Scandal in Bohemia). He also refers to the knowledge obtained from previous cases (e.g. The Veiled Lodger), that is, from gained experience as well as from *trained observation* (The Empty House). All these procedures have to come together, and it is in The Five Orange Pips that Holmes clearly states his philosophy on the art of investigation: *"The ideal reasoner would, when he has once been shown a single fact in all its bearings, deduce from it not only all the chain of events which led up to it, but also all the results which would follow from it."*

Holmes then goes on to consider the importance of knowledge and states his well-held views: *"Problems may be solved in the study which have baffled all those who have sought a solution by the aid of their senses. To carry the art, however to its highest pitch, it is necessary that the reasoner should be able to utilize all the facts which have come to his knowledge; and this in itself implies, as you will readily see, a possession of all knowledge, which, even in these days of free education and encyclopedias, is a somewhat rare accomplishment."*

SECOND METHOD OF DEDUCTION

The second method of deduction is really deduction using science in all its aspects. This branch of detective work is the domain of the forensic scientist, and Holmes was an early experimenter in this field [1B].

He tells Watson that his 'methods' became known whilst he was still at university, where he first started investigating cases. When he came to London he studied *"all those branches of science which might make me more efficient"* (The Musgrave Ritual).

He also refers to an occasion when Watson, on his behalf, scrutinized his qualifications, accomplishments and even addictions. Most of the former appear to be quite impressive, with the exception of philosophy, astronomy and politics, at which it seems Holmes was quite hopeless. This is strange because surely philosophy is essential for a reasoner?

However, ignoring his addictive habits, the report by Watson does get somewhat better: *"Botany variable, geology profound, chemistry eccentric, anatomy unsystematic, sensational literature, crime records unique, violin player, boxer, swordsman, lawyer, and self-poisoner by cocaine and tobacco. Those, I think, were the main points of my analysis."* (The Five Orange Pips).

From the above, it is evident that Holmes places great store on scientific knowledge as a means of assisting with the effective solving of his cases. This knowledge is necessarily at the heart of any type of deduction that requires scientific analysis using tests or microscopic examination in the laboratory. Similar methods are employed by present day forensic scientists, who of course have access to modern equipment and recent techniques such as DNA[1B] testing.

THE LABORATORY

Deduction using science inevitably requires a laboratory, and some of the stories reflect this with their descriptive accounts of Holmes carrying out various scientific procedures at his bench.

In fact when Watson first meets Holmes, he is busy working at the bench in the chemical laboratory of St. Bartholomew's Hospital [1B]. However before this meeting, Stamford, Watson's friend, describes Holmes to him in some detail, telling him that: *"He is a little queer in his ideas – an enthusiast in some branches of science."* (A Study in Scarlet). Stamford also mentions that Holmes is good at anatomy and chemistry, and although *"his studies are very desultory and eccentric, he has amassed a lot of out-of-the-way knowledge which would astonish his professors".*

Watson is then given more information about Holmes on their way to meet him. Stamford considers him to be *"a little too scientific for my tastes."* He tells Watson that Holmes has a cold-blooded streak, and would not be past giving a vegetable alkaloid [1B] to a friend to find out the effect, although, to be fair, he would also probably be willing to take it himself. *"He appears to have a passion for definite and exact knowledge,"* he adds, and then proceeds to describe how Holmes on one occasion was seen to beat specimens in the dissecting room with a stick. The explanation for this bizarre behaviour was *"to verify how far bruises may be produced after death".*

The Valley of Fear has Watson describing the look on Holmes' face when he learns of the murder of Jack (John) Douglas. *There was no trace then of the horror, but his face showed rather the quiet and interested composure of the chemist who sees the crystals falling into*

position from his over-saturated solution [1C].

A brief mention has already been made of Holmes' 'laboratory', which is basically a bench, in the Baker Street consulting room. This is where Holmes carries out his scientific and chemical experiments. They are sometimes malodorous (as Watson puts it), creating smells to the discomfort of both the good doctor and their faithful housekeeper, Mrs. Hudson. At the start of **The Mazarin Stone**, Watson surveys the scene in the untidy but familiar first floor consulting room at Baker Street, and likes what he sees: *He looked round him at the scientific charts* [1C] *upon the wall, the acid-charred* [1C] *bench of chemicals, the violin-case leaning in the corner, the coal-scuttle* [1B]*, which contained of old the pipes and tobacco.*

A clear description, but this raises the question as to what Holmes used for the coal? He could have used his violin case, but from the above excerpt it seems an unlikely receptacle, unless it needed filling at the time.

HOLMES' PUBLICATIONS

Throughout the canon there are references to Holmes' monographs and published works. His detailed monographs are on unusual topics, of which very few and probably none, would have reached the bestseller charts of the day, if there had been any. Holmes wrote several of these monographs, and some of the scientific and technical ones are mentioned in other parts of this book, so only brief comment is made here. A few are listed in **The Sign of Four**, namely the ones on 'Tobacco Ashes' (also mentioned in **A Study in Scarlet**), the 'Tracing of Footsteps', and 'The Effect of Trades on the Form of the Human Hand'. Holmes' proposed volume on 'Detection', which surely would have been his magnum opus, is mentioned in **The Abbey Grange**. *"I propose to devote my declining years to the composition of a textbook, which shall focus the whole art of detection into one volume."*

We know he also wrote magazine articles, such as 'The Book of Life' (**A Study in Scarlet**) with its rather misleading title, as well as monographs on the following subjects: 'The Human Ear' (The Cardboard Box); 'The Dating of Old Documents' (The Hound of the Baskervilles); 'Tattoo Marks' (The Red-Headed League); 'Secret Writings' (The Dancing Men); 'The Use of Dogs in Detective Work' (The Creeping Man); 'Bee Culture' (His Last Bow); and 'The Polyphonic Motets of Lassus'[1C] (The Bruce-Partington Plans): and that he thought of writing ones on 'Malingering' (The Dying Detective) and 'The Typewriter and Its Relation to Crime' (A Case of Identity).

FILM, THEATRE AND OTHER PRODUCTIONS

Finally, it is obvious that the popularity of the Sherlock Holmes stories and novels will continue unabated for many years. Over time their success has increased as a result of the many screen, television, radio and theatre versions that have been produced. These, augmented by video and other recorded formats, have created a world-wide audience for the adventures. In this way the character of Holmes, as portrayed over the years by actors such as John Barrymore, Jeremy Brett, Peter Cushing, Carleton Hobbs, Ian Richardson and Basil Rathbone, has been brought to life. These productions, with excellent performances from leading actors and good supporting casts, have provided first rate entertainment to please the many enthusiastic fans of the cases.

This great collection of adventures and novels which we are so fortunate to have, should undoubtedly be regarded as a literary treasure. The cases of Sherlock Holmes as recorded by Doctor Watson are a detailed record of the accomplishments of the great detective, and Watson's own views on these are made clear as shown in the last two quotes of this chapter. On one occasion when admiring Holmes' skill at disguise (A Scandal in Bohemia), Watson points out that: *The stage lost a fine actor, even as science lost an acute reasoner, when he became a specialist in crime.*

Another time he feels impelled to praise Holmes' methods. In A Study in Scarlet he gives the master detective a magnificent accolade, which in itself provides a fitting conclusion to this introduction. *"You have brought detection as near an exact science as it ever will be brought in this world."*

Watson and Holmes

NOTES 1A

Ulster: A long overcoat made of coarse cloth, and sometimes worn with a belt. It originated in Belfast (Ulster) where the cloth was made.

Deerstalker: A double-peaked cloth cap, shaped to the head, with flaps tied on top.

The Strand Magazine was started by publisher George Newnes in 1891 and continued until 1950. Its heyday was in the Victorian and Edwardian era when it had a large circulation with readers attracted by its excellent short stories. Other notable authors of these included H.G. Wells and P.G. Wodehouse.

The **Hansom cab** was a favourite form of transport for the better-off Victorians. It was designed in 1834 by the architect Joseph Aloysius Hansom, from whom it obtained its name, but it was later greatly improved by John Chapman who, with W. S. Gillet patented his design in 1836. A precursor of the modern London taxi cab, it was a one-horse vehicle, with the driver perched high at the back (although the early version had the driver at the front) and with one very large wheel on each side. The passenger entered via a platform and through a pair of padded half-doors at the front of the cab.

The **Growler, or Clarence cab**, was a four-wheeled horse-drawn vehicle in use at the time. The popular design of this cab dates from the 1830s (though some four-wheeled cabs were in use in the 1820s), and required one or two horses to pull the vehicle. It usually carried four passengers. The name Growler was a London slang term given to the vehicles, either because of the noise they made, or because of their ubiquitous, surly drivers and slow horses. Sometimes though there were probably slow drivers and surly horses if the oats weren't up to scratch.

Titles of stories in which **Holmes fails** in some respect, though not always totally: The Dancing Men, A Scandal in Bohemia, The Five Orange Pips, The Yellow Face.

NOTES 1B

Coal-scuttle: Container, usually of metal but sometimes of wood, which held coal for the open fire heating the room. References to the **coal-scuttle** are found in The Mazarin Stone, The Musgrave Ritual.

A Persian slipper is a soft shoe which can easily be slipped on or off, and has a curled end at the toe. References to the **Persian slipper** are found in The Illustrious Client, The Empty House, The Musgrave Ritual and The Naval Treaty

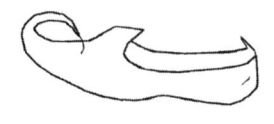

Professor Moriarty. It has been suggested that the mathematical abilities of the Professor were based on those of George Boole (inventor of Boolean algebra). In 2003 there was a further suggestion that the name of the Professor may have been obtained by Conan Doyle from a member of the Sheringham Golf Club in Norfolk, of which it is said that Doyle himself was once a member. The major story about **Professor Moriarty** is The Final Problem and he is also mentioned in The Empty House. Several other adventures have some connection with the master criminal or references to him.

Sherlock Holmes, FRSC (Hon). In October 2002 the Royal Society of Chemistry announced that it was awarding an Extraordinary Honorary Fellowship to Sherlock Holmes for his pioneering work in using chemistry in **Forensic Science** for the detection and solving of crime.

DNA: Deoxyribonucleic acid carries the genetic code of an individual. DNA profiling, used in forensic science, was introduced as a result of the work of Sir Alec Jeffreys, a pioneer in this field in the 1980s. (See Note 6B)

Saint Bartholomew's Hospital (popularly known as Bart's), at Smithfield in the City of London, is one of the oldest hospitals in Britain and was founded by Rahere, an Augustinian monk, in 1123.

Vegetable alkaloids are substances obtained from plants. They include strong poisons, drugs and antidotes. Strychnine is an example of a powerful and dangerous vegetable alkaloid.

NOTES 1C

Baker Street Irregulars: Holmes employs a band of poorly clad but energetic young boys who act as scouts or information gatherers. Their pay is one shilling (5 pence) a day, with a bonus for the boy who first comes up with a result. Their small size and dress allows them to penetrate all parts of the metropolis comparatively unnoticed. Wiggins is the leader of the group in A Study in Scarlet and The Sign of Four, and Simpson keeps watch in The Crooked Man.

Oversaturated Solution and Crystallization refers to a method of chemical purification whereby a warm or hot super-saturated solution of certain salts is allowed to cool, causing crystals of the salt to separate out and solidify.

Baker Street Irregulars

Scientific charts: We can only conjecture what these may have been. Almost certainly there would have been one displaying the Periodic Table of the elements, showing atomic numbers and atomic weights. (The Periodic table was first suggested by Dmitri Mendeleev in 1869). Another chart may have shown the structure of an atom, and there may well have been a table of compounds giving melting points, specific gravities (relative densities) and specific heats of substances and other chemical constants. Perhaps also there was a chart showing the indications of various chemical tests with litmus paper and reagents such as methyl orange and phenolphthalein. Holmes could also have used charts to show the results of his own experiments, and Watson may well have contributed anatomical diagrams to adorn the walls, although these would probably have upset Mrs. Hudson and some of Holmes' visitors, and so are less likely to have been displayed.

Acid charred bench: Some acids, usually concentrated, are highly corrosive and will burn or score marks in surfaces such as wood.

Polyphonic motets of Lassus: Lassus was a 16th century Flemish composer of choral music. The motets are vocal pieces with harmonized melodies.

2 THE MEDICAL RECORD

MEDICINE AND PHARMACOLOGY

After qualifying as a doctor Arthur Conan Doyle practised in Southsea and also in London. To make ends meet, and as patients were few and far between (there was no National Health Service in those days), he turned to writing. In his detective stories, which were to become so successful, Conan Doyle made Watson an ex-army doctor, as well as the recorder and narrator of many of the cases (with a few exceptions [2A]) on which Holmes worked.

To be successful, the important principle of any writer is to write about what he or she knows and understands. As a doctor himself, it was both practicable and easy for Conan Doyle to use his own medical knowledge to add interest and authenticity to Watson's accounts. Watson invariably accompanies Holmes on his numerous investigations, and as a competent medical practitioner, diagnoses and sometimes treats the many medical conditions and injuries mentioned in the stories.

At the beginning of A Study in Scarlet Doctor Watson tells us something about himself. Having obtained his medical qualification at London University in 1878, Watson went on to Netley [2A] to train as an army surgeon. Here he completed his studies and then became assistant surgeon to the Fifth Northumberland Fusiliers [2A], which were stationed in India. At this time army surgery would have been a particularly gory process on the battlefield, and facilities would have been basic to say the least.

Later, Watson joined the Berkshires [2A], and was wounded by a Jezail [2A] bullet, which shattered the bone and grazed the subclavian artery [2A], at the battle of Maiwand [2A] in the Afghan Wars. He was hospitalised at Peshawar [2A], and although gradually improved in health, was then *struck down by enteric fever* [2A], *that curse of our Indian possessions* [2B]. As a result he became weak and emaciated, as he tells us, and was sent back to England to recover (A Study in Scarlet). One day at the Criterion Bar in London, Watson, who previously practised at St Bartholomew's

Hospital [2B], bumps into 'young Stamford', who had been his dresser [2B] at Bart's. Stamford takes him to meet Sherlock Holmes, and so the association with the famous detective begins.

As a close companion of Holmes, Watson is sometimes put in danger by his proximity to the detective when confrontations with criminals occur. On one such occasion Holmes shows great concern when Watson is shot, but luckily he only suffers a superficial wound (The Three Garridebs).

DOCTOR WATSON'S MEDICAL PRACTICE

In A Scandal in Bohemia Holmes deduces that Watson has started practising again as a doctor by noticing *"a bulge on the side of his top hat to show where he has secreted his stethoscope* [2B]*."* A hat may be an unusual though convenient position for the instrument, but this is a part of the apparel that would easily gather dirt in the smoke-laden atmosphere of the times, and so endanger the hygienic cleanliness of the stethoscope when used on a patient.

In some of the cases, Watson obtains a stand-in (or locum) for his medical practice when Holmes desires his company in an investigation. One of these is Anstruther in The Boscombe Valley Mystery; while in The Crooked Man he tells Holmes *"I have no doubt Jackson would take my practice"*.

In two other cases (The Final Problem; The Stockbroker's Clerk) the replacement doctors are unnamed. Watson tells Holmes *"I have an accommodating neighbour,"* and *"I do my neighbour's when he goes. He is always ready to work off the debt."*

One day when standing on Watson's doorstep, Holmes makes an interesting observation about his neighbour's practice. *"Your neighbour is a doctor"* said he, nodding at the brass plate.
"Yes. He bought a practice as I did."
"An old-established one?"
"Just the same as mine. Both have been ever since the houses were built."

"Ah! then you got hold of the best of the two."
"I think I did. But how do you know?"
"By the steps, my boy. Yours are worn three inches deeper than his."
(The Stockbroker's Clerk).

In the same story the reader is told that Watson purchased his medical practice in Paddington[2B] from an elderly physician: *Old Mr. Farquhar, from whom I purchased it, had at one time an excellent general practice; but his age, and an affliction of the nature of St. Vitus' dance*[2B] *from which he suffered, had very much thinned it.*

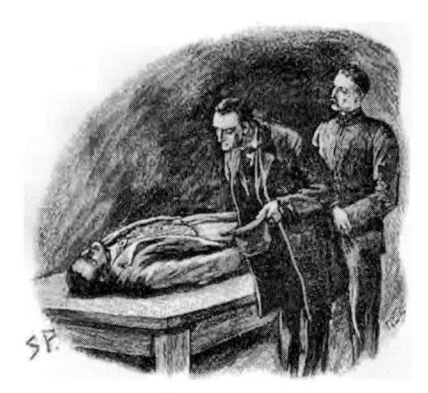

The Norwood Builder account informs us that Watson sold his Kensington medical practice to a Doctor Verner, a distant relation of Holmes. Watson only learns of this last fact some years later, and that Holmes himself put up the money for Verner to buy it.

OTHER MEDICAL PRACTITIONERS

The Speckled Band has an interesting analogy with recent events. Here Holmes remarks that: *"When a doctor does go wrong he is the first of criminals. He has nerve and he has knowledge."* This is as true today as it was then. Holmes then mentions two medical men, Palmer and Pritchard [2B], who became criminals.

Medics can also be on the receiving end of a crime, as shown in The Retired Colourman, where a young doctor, Ray Ernest, becomes a murder victim when he establishes a relationship with Mrs. Amberley.

In The Dying Detective an ailing Holmes tells Watson that having carried out some recent medico-criminal research and learnt something of Eastern diseases: *"I contracted this complaint."* Watson who wants to treat him, is understandably hurt when he is rebuffed, with Holmes telling him that: *"You are only a general practitioner with very limited experience and mediocre qualifications."* Therefore Watson suggests: *"Let me bring Sir Jasper Meek or Penrose Fisher, or any of the best men in London."* He then offers to fetch Doctor Ainstree: *the greatest living authority upon tropical disease.* But the great detective has his own idea of whom he will see. Instead he sends Watson to fetch Mr. Culverton Smith, a planter who has a special knowledge of the disease that Holmes is suffering from. Watson manages to find Culverton Smith, who points out that he is only an amateur expert on disease, indicating *a row of bottles and jars which stood upon a side table.* These containers hold gelatine cultures [2B] nurturing the various diseases that Smith studies.

Mr Kent is the physician who attends Godfrey Emsworth, a sufferer from suspected leprosy. When Holmes is brought in to investigate Godfrey's 'disappearance', he realizes that a medical solution to the mystery is required, and so brings Sir James Saunders, a specialist in leprosy and skin diseases, to look at Emsworth (The Blanched Soldier).

In 1897 Watson reports that Holmes' heavy workload is taking its toll. *In March of that year Doctor Moore Agar, of Harley Street*[2B]*, gave positive injunctions that the famous private agent lay aside all his cases, and surrender himself to complete rest if he wished to avert an absolute breakdown* (The Devil's Foot).

A nervous little doctor inadvertently sets off the mutiny on the convict ship named the *Gloria Scott*, by finding a pair of pistols at the foot of a prisoner's bed. Later he comes to a sticky end himself when attacked by the rebel ringleader. *Prendergast with his own hands cut the throat of the unfortunate surgeon* (The Gloria Scott).

Finally two famous London streets, one of notable medical importance, are linked in another story: *"This is Baker Street, not Harley Street* [2B] *",* says Holmes when John Mason states bluntly that he thinks his employer, Sir Robert, has gone mad (Shoscombe Old Place).

HOSPITALS

Doctor Trevelyan tells Holmes and Watson (The Resident Patient) that he was a researcher, occupying a minor position in King's College Hospital [2B].

On an occasion when Holmes is injured after being set upon by two ruffians (The Illustrious Client), he is taken to Charing Cross Hospital [2C] and attended by the famous surgeon, Sir Leslie Oakshott. This is also the hospital in which Doctor Mortimer (The Hound of the Baskervilles) worked as a house surgeon. *From his friends of the C.C.H.* is found engraved on his walking stick, when he leaves it behind at 221B Baker Street: C.C.H. of course standing for Charing Cross Hospital.

St Bartholomew's Hospital is mentioned in other parts of this book.

POISONS

When on his deathbed, a bottle containing the bitter tasting quinine [2C] is mentioned by Major John Sholto to his sons Bartholomew and Thaddeus. The latter son is a hypochondriac and asks Watson about various *quack nostrums* [2C], some of which he bore about in a leather case in his pocket.

Watson is upset on learning that Miss Morstan, whom he loves, is to become a wealthy woman and will therefore be unlikely to accept him as her suitor, or so he thinks. With his mind elsewhere, he gives an inaccurate and dangerous answer to Thaddeus's medical questions. *Holmes declares that he overheard me caution him* [Thaddeus] *against the great danger of taking more than two drops of castor oil* [2C], *while I recommended strychnine* [2C] *in large doses as a sedative.* As we can see from Watson's own account, Conan Doyle wrote these words with some humour in mind.

Ironically in the same case, Watson actually has to confront strychnine poisoning when the other brother dies. *"Death from some strychnine-like substance which would produce tetanus* [2C]*",* explains Watson on looking at the body of Bartholomew Sholto, and replying to Holmes' query as to the cause of death. The detective has already commented on the *"distortion of the face, this Hippocratic smile* [2C], *or 'risus sardonicus', as the old writers called it"* (The Sign of Four). It turns out that the substance has been administered by a poisoned long dark thorn in the skin just above the ear. This tipped thorn has been blown from a blow-pipe, and it is discovered that Tonga, a short Andaman [2C] islander and companion of Jonathan Small, was the pipe blower.

Later Holmes finds a small woven grass pouch about the size of a cigarette-case. *Inside were half a dozen spines of dark wood, sharp at one end and rounded at the other, like that which had struck Bartholomew Sholto.* When Watson examines one of these thorns he finds that *it was long, sharp, and black, with a glazed look near the point as though some gummy substance had dried upon it. The blunt end had been trimmed and rounded off with a knife* (The Sign of Four).

Small himself was a prisoner at the penal colony on Blair Island [2C], and later explains that the area *was infested with wild cannibal natives, who were ready enough to blow a poisoned dart at us if they saw a chance.* While there, Small became a dispenser of drugs. *"I learned to dispense drugs for the surgeon, and picked up a smattering of his knowledge."*

As Holmes discovers, the murder of Mr. Enoch J. Drebber (A Study in Scarlet) is carried out by using poison. *"Having sniffed the dead man's lips I detected a slightly sour smell, and I came to the conclusion that he had had poison forced upon him",* Holmes tells Watson, pointing out that the similar cases *"of Dolsky in Odessa, and of Leturier in Montpellier will occur at once to any toxicologist* [2D]*".*

In the same story the murderer, Jefferson Hope, describes how he obtained and used South American arrow poison [2D] to make a pill with which to kill Enoch Drebber. *"Among the many billets which I have filled in America during my wandering life, I was once janitor and sweeper-out of the laboratory at York College. One day the professor was lecturing on poisons, and he showed his students some alkaloid, as he called it, which he had extracted from South American arrow poison, and which was so powerful that the least grain meant instant death. I spotted the bottle in which this preparation was kept, and when they were all gone, I helped myself to a little of it. I was a fairly good dispenser, so I worked this alkaloid into small, soluble pills, and each pill I put in a box with a similar pill made without the poison."*

Another case (The Sussex Vampire) also centres on the use of a South American poison, and here Holmes describes the consequences of administering it to a youngster. *"If the child were pricked with one of those arrows dipped in curare* [2D] *or some other devilish drug, it would mean death if the venom were not sucked out."* He also refers to the family pet on which the drug is first tried: *"And the dog! If one were to use such a poison, would one not try it first in order to see that it had not lost its power?"* Fortunately the child's life is saved by sucking out the venom, and Holmes declares: *"Was there not a queen in English history who sucked such a wound to draw poison from it?"* However he does not supply the answer [2D].

Mrs. Ronder considers committing suicide using poison, but after Holmes deals with her case satisfactorily (The Veiled Lodger), she sends him a small blue bottle with a red poison label. This contains prussic acid [2D], the substance with which she intended to kill herself. Holmes is pleased to receive it because he knows that the chemical is now out of her reach and temptation.

In another story (The Devil's Foot) Doctor Sterndale supplies a good account of the evil Devil's-foot root [2D] *"It has not yet found its way either into the pharmacopoeia* [2D] *or into the literature of toxicology. The root is shaped like a foot, half human, half goatlike; hence the fanciful name given by a botanical missionary. It is used as an ordeal poison* [2D] *by the medicine-men in certain districts of West Africa, and is kept as a secret among them. This particular specimen I obtained under very extraordinary circumstances in the Ubanghi country* [2D].*"* He opened the paper as he spoke, and disclosed a heap of reddish-brown, snuff-like powder.

When Holmes and Watson experiment with this powder, we as readers learn of its horrific effect in a striking account that effectively describes the feeling of ill-boding which comes over them both. *"Well, then, I take our powder–or what remains of it–from the envelope, and I lay it above the burning lamp. So! Now, Watson, let us sit down and await developments." They were not long in coming. I had hardly settled in my chair before I was conscious of a thick, musky odour, subtle and nauseous. At the very first whiff of it my brain and my imagination were beyond all control. A thick, black cloud swirled before my eyes, and my mind told me that in this cloud, unseen as yet, but about to spring out upon my appalled senses, lurked all that was vaguely horrible, all that was monstrous and inconceivably wicked in the universe. Vague shapes swirled and swam amid the dark cloud-bank, each a menace and a warning of something coming, the advent of some unspeakable dweller upon the threshold, whose very shadow would blast my soul.*

Poison pills are discovered in A Study in Scarlet when Lestrade produces a small chip [2D] ointment box containing a couple of pills. This box is lying on the window-sill of the room in which Joseph Stangerson is found murdered. On Holmes' request, Lestrade hands over the box, and Holmes asks Watson for his opinion on the contents. Watson obliges. The pills are of a pearly grey colour, small, round, and almost transparent against the light. Watson thinks they are soluble in water, and so Holmes dissolves half of one of the pills in water and gives it to the landlady's dog, that is already dying but suffering in the process. When nothing happens, Holmes shows his chagrin and disappointment. He then tries the second pill in the same way, halving it and giving one of the halves to the poor dog, which dies quickly and peacefully as intended. Holmes naturally concludes: *"Of the two pills in that box, one was of the most deadly poison, and the other was entirely harmless."*

Another box, this time containing a concealed poisoned spring, is sent to Holmes by Mr. Culverton Smith (The Dying Detective). Holmes is wary of packages in his post and does not open the small ivory box. The poison is unspecified but is suspected to consist of the germs of an Asiatic disease, which has already killed Victor Savage, Smith's nephew.

A self-inflicted poisoning is successful at the end of another case (The Golden Pince-Nez) when Anna Coram consumes a poisonous substance. Holmes rushes forward to remove the phial from her hand, but she shouts out *"Too late! I took the poison before I left my hiding-place."*

Similarly when Holmes corners the murderer in The Retired Colourman, the man tries to take poison but *Holmes sprang at his throat like a tiger and twisted his face towards the ground. A white pellet fell from between his gasping lips.* He cries. *"No short cuts, Josiah Amberley. Things must be done decently and in order."* However we are not told what the poisoned pellet contained, though probably it was a fast-acting cyanide pill.

COCAINE

By no means all the drugs mentioned in the canon are used for medicinal purposes. Holmes himself is addicted and participates in drug-abuse at times. He likes to inject himself with cocaine [2E] (The Sign of Four). A description of his method using a hypodermic syringe opens this adventure, and here we learn that Holmes has been addicted for some time, *as his sinewy forearm and wrist was all dotted and scarred with innumerable puncture-marks.* Watson, both as his friend and as a medical man, of course thoroughly disapproves of this habit: *Three times a day for many months I had witnessed this performance, but custom had not reconciled my mind to it, and my conscience swelled nightly within me that I had lacked the courage to protest.* Holmes on this occasion is taking a 7% solution of cocaine. Watson enquires if the drug is morphine or cocaine, and bridles when Holmes asks if he would like to try it. He urges Holmes to *"count the cost!"* and goes on to emphasize the effect the drug could have by causing *"increased tissue-change, and may at least leave a permanent weakness."*

Watson then sets Holmes a short deduction test by handing him his brother's watch. Holmes responds by remarking that: *"it would prevent me from taking a second dose of cocaine".* The test is a way of stimulating his mind as an alternative to the drug, and so Watson questions him: *"May I ask whether you have any professional inquiry on foot at present?"* To which Holmes answers: *"None. Hence the cocaine. I cannot live without brainwork."* But, for all Watson's efforts to get Holmes away from drugs, at the very end of the case it seems that Holmes is unreformed, as he says *"For me, there still remains the cocaine-bottle."* *And he stretched his long white hand up for it.*

In another adventure (A Scandal in Bohemia) Watson reports that Holmes was *"alternating from week to week between cocaine and ambition, the drowsiness of the drug, and the fierce energy of his own keen nature"*. But in March 1888, Watson finds him *"risen out of his drug-created dreams, and hot upon the scent of some new problem"*.

The Dying Detective contains a short passage describing the mantelpiece in the consulting room with its litter of items including syringes, thus emphasizing Holmes' addiction.

OPIUM, MORPHINE, MORPHIA

When Isa Whitney, a friend of Watson's, becomes addicted to opium [2E], Watson goes to look for him in the East End of London, and surprisingly discovers both his friend and Holmes in the same opium den (The Man with the Twisted Lip).

The story opens by explaining that Whitney has become addicted after reading Thomas De Quincey's book [2E] describing the dreams and sensations under the influence of opium. Whitney in imitation *had drenched his tobacco with laudanum* [2E] *in an attempt to produce the same effects*. As is only too well known today, Watson mentions that the habit is easier to start than to get rid of, and proceeds to describe Whitney's appearance under the influence: *"I can see him now, with yellow, pasty face, drooping lids and pin-point pupils, all huddled in a chair, the wreck and ruin of a noble man"*.

When Watson enters the opium den, the air is thick and heavy with the brown opium smoke. There then follows an evocative description of the interior and of the attitudes of the smokers as the burning poison waxed or waned [2E] in the bowls of the metal pipes. He finds and rescues his friend, but to his amazement he also finds Holmes there in the disguise of an old man, with an opium pipe dangling down from between his knees.

Opium also features in two cases (Silver Blaze, and Wisteria Lodge) where the drug is administered without the consent of the persons involved. In the first, Ned Hunter, a stable lad, is drugged when powdered opium is added to his meal of curried mutton. In the second,

Miss Burnet is also drugged with opium, which is placed in her lunch. On release from her imprisonment and ordeal, she is given two cups of strong coffee to clear her brain from the mists of the drug. Also when Jonathan Small (The Sign of Four) was on guard duty in Agra, he tells us that the rebels would become drunk with opium and with bang [2E].

"Oil [2F], opium, morphia [2F]!" he cried. "Anything to ease this infernal agony" is the urgent request made by Ian Murdoch as he bursts in on Holmes and Inspector Bardle (The Lion's Mane). They can both see the fierce red patterned weal on Murdoch's shoulder which is giving him great pain, but he also has some much more worrying symptoms as this account describes: *"for the sufferer's breathing would stop for a time, his face would turn black, and then with loud gasps he would clap his hand to his heart, while his brow dropped beads of sweat. At any moment he might die".* Fortunately he recovers to health.

A situation in which Watson administers morphia occurs in The Illustrious Client. Baron Gruner has had vitriol thrown in his face, and Watson who is present, records that: *"I seized a carafe [2F] from a side-table and rushed to his aid".* After explaining the attack to the servants in the house, Watson continues his treatment of the victim: *"I bathed his face in oil [2F], put cotton wadding on the raw surfaces, and administered a hypodermic of morphia. I was relieved when his family surgeon, closely followed by a specialist [2F], came to relieve me of my charge."*

Holmes also receives morphine after a vicious attack in the same story, which precedes the episode above.

POTIONS AND SCENTS

When Mrs. Hudson, the housekeeper at 221B Baker Street, is worried about Holmes' health one day, she tells Watson that she suggested a cooling medicine [2F] to help him, *"but he turned on me sir, with such a look that I don't know how ever I got out of the room"* (The Sign of Four).

This story also has a number of other references. One aromatic substance which can create a calming or soothing effect on those who encounter it is incense [2F]. This odoriferous product is burning at the home of Thaddeus Sholto when Holmes, Watson and Miss Morstan enter the house. *A lamp in the fashion of a silver dove was hung from an almost invisible golden wire in the centre of the room. As it burned it filled the air with a subtle and aromatic odour.*

The Creeping Man describes the use of a potion extracted from a monkey in an attempt to obtain rejuvenation and longevity. When Holmes manages to open Professor Presbury's box, which has been guarded so fiercely by the man himself, he finds an empty phial, another nearly full, and a hypodermic syringe. A little-known scientist named Lowenstein in Prague [2F], is supplying through a London agent, an extract from the glands of the black-faced langur monkey [2F]. This extract is intended to be an elixir of life, but the Professor's use of the serum has devastating effects, and Lowenstein in a letter to him urges caution, and suggests that the serum from an anthropoid [2F] creature which walks erect and is in all ways nearer to man would be better.

DISEASES

A highly infectious disease is mentioned in A Study in Scarlet. A policeman explains to Holmes that two houses are empty because the landlord *"won't have the drains seed to, though the very last tenant what lived in one of them died o' typhoid [2F] fever."* Note the excruciating language!

Another disease which was a scourge in Victorian and Edwardian times, was consumption [2F] or tuberculosis. Godfrey Staunton's wife suffered and died from a particularly virulent form of the disease (The Missing Three-Quarter), and a tutor named Fraser was also consumptive, and died from the same complaint (The Hound of the Baskervilles). Fraser had been a tutor at the school set up in Yorkshire by Stapleton.

INJURIES

A great variety of injuries, some of which result in death, are documented in the stories. They are so numerous that it is impossible to mention them all and only a small number can be included here.

For example in **The Six Napoleons** Pietro Venucci dies of a clasp knife [2G] wound with *"a great gash in his throat"*, and another man to die from a knife wound is Eduardo Lucas, murdered by means of an Indian dagger (**The Second Stain**). Another severe attack which causes death can be found in **Black Peter**, where a terrible injury is inflicted by means of a harpoon, leading Holmes to experiment with a barbed spear and a pig's carcass in order to find how easy it would be to kill with a single thrust of this weapon.

In **The Stockbroker's Clerk**, Watson (who is sometimes called upon to examine, diagnose or treat a victim or sufferer) has to revive a Mr. Pinner, who has attempted suicide. *"I stooped over him and examined him. His pulse was feeble and intermittent, but his breathing grew longer, and there was a little shivering of his eyelids, which showed a thin white slit of ball beneath."* An opened window to provide air, and cold water poured on Mr. Pinner's face, lead to restoration of life. Although Pinner is a criminal, the doctor performs his professional duty well, as the calling of his profession demands.

In another case (**The Illustrious Client**) Holmes is a victim, and suffers injuries after a vicious attack. His surgeon, Sir Leslie Oakshott, reports on the extent of these as follows: *"Two lacerated scalp wounds and* *some considerable bruises. Several stitches* [2G] *have been necessary."* With permission Watson enters the room where Holmes lies with his head bandaged. A crimson patch had soaked through the white linen compress [2G]. *Holmes then mutters in a weak voice: "All right, Watson. Don't look so scared. It's not as bad as it seems."*

MEDICAL PROBLEMS

An aortic aneurysm [2G] is mentioned in A Study in Scarlet. Jefferson Hope has this condition, as Watson detects: *"on placing his hand on the man's chest, becoming at once conscious of an extraordinary throbbing and commotion which was going on inside."*

In the same adventure, Holmes in his investigation concludes that *"the blood which covered the floor had burst from the murderer's nose* [2G] *in his excitement".* Because of this, Holmes hazards *"the opinion that the criminal was probably a robust and ruddy-faced man."*

Watson observes that John Turner is in the grip of some deadly and chronic disease (The Boscombe Valley Mystery), whilst Turner himself admits that he is a dying man, and tells Watson: *"I have had diabetes* [2G] *for years. My doctor says it is a question whether I shall live a month."*

Doctor Barnicot is a medical practitioner with premises in Kennington Road and also at Lower Brixton Road. He owns two busts of Napoleon which are individually removed and destroyed at each of his premises. Watson suggests that the person who did this may have been suffering from monomania [2G] and had an 'idée fixe' about Napoleon (The Six Napoleons).

Another medical practitioner, Doctor Trevelyan (The Resident Patient), calls to consult Holmes, who notices the brougham at the door, and inside it the doctor's wicker basket containing his medical instruments. Watson recognizes the doctor's name on introduction, and asks him: *"Are you not the author of a monograph upon obscure nervous lesions* [2G]*?"*

Later in the adventure Dr. Trevelyan relates how he intended to treat a patient he thought was suffering from an attack of catalepsy [2G]: *"I had obtained good results in such cases by the inhalation of nitrite of amyl [2G]".* Holmes tells Watson that in the past he has imitated the symptoms of catalepsy, the condition used by criminal perpetrators to gain access to the premises of Doctor Trevelyan. As the story continues, a further diagnosis is required: *"I should say he has been dead about three hours, judging by the rigidity of the muscles [2G],"* states Watson on looking at the body of Blessington.

This adventure with its high medical content must have given Conan Doyle a great deal of satisfaction to write.

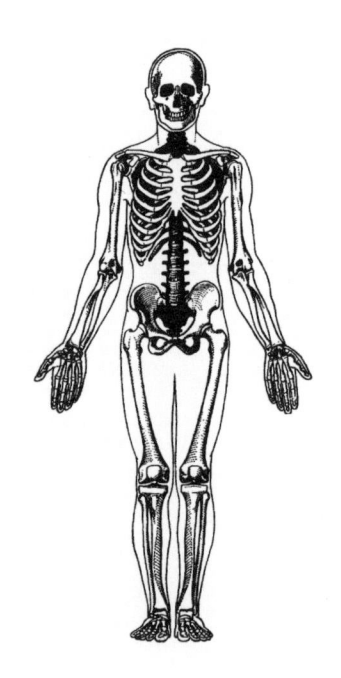

NOTES 2A

Stories where Holmes is the narrator: Holmes writes up the accounts of The Lion's Mane, and The Blanched Soldier. He narrates details of The Musgrave Ritual to Watson.

At Netley, near Southampton, there was a huge military hospital to cope with a thousand soldiers wounded in the Crimean War. The size of the hospital was striking, being over a quarter of a mile in length. With so many wounded and ill soldiers, this was obviously an ideal training establishment for army surgeons.

The Fusiliers: The name originated from the weapon (the fusil – a light flintlock musket) used by these special troops, whose regiments date from the 17th century.

The Berkshires (or more accurately The Royal Berkshire Regiment) was formed from an amalgamation in 1881 of two other regiments.

The Jezail bullet was the one used in the Afghan musket of that name.

The subclavian arteries, of which there are two, supply the neck and arms with blood.

Maiwand, in Afghanistan, was the scene of a British defeat on 27 July 1880. General Burrows was defending Kandahar and marched to meet the Afghans under Ayub Khan at Maiwand, but had to retreat with considerable British losses.

Peshawar: An important trade centre at the eastern end of the Khyber Pass. It was situated in India when the story was published, but is now in Pakistan.

Enteric fever is an infection causing intestinal inflammation and includes the serious diseases of paratyphoid and typhoid fever. Paratyphoid fever is caused by *Salmonella paratyphi* A, B and C, and typhoid fever is caused by *Salmonella typhi*. Infection from contaminated food and water, and from those already suffering from the disease are the means by which it spreads.

NOTES 2B

Indian possessions: India was part of the British Empire until independence in 1947.

Saint Bartholomew's Hospital (popularly known as Bart's**)** in the City of London is one of the oldest hospitals in Britain, and was founded by Rahere, an Augustinian monk, in 1123.

Dresser: A surgeon's assistant who dressed wounds and carried out other medical tasks.

The **Stethoscope** is a medical instrument used by doctors to listen to the heart and lungs.

Paddington: A district of London close to Baker Street.

St. Vitus' dance: Also known as Sydenham's chorea after the 17th century English physician Thomas Sydenham. Uncontrollable twitching and jerky movements, confusion and emotional disturbance are symptoms.

William Palmer (1824-56) poisoned a friend, and probably his wife, and brother as well. **Edward William Pritchard (1825-65)** was an English surgeon who poisoned his wife and mother-in-law. Their crimes, although very serious, in no way compare with those of Doctor Harold Shipman, who was jailed in 2000 for a whole series of murders and died in prison in 2004. [Details of the Palmer and Pritchard cases can be found in *Medical Murders*, Goodman, J., Ed., 1991. There are also recent books about Doctor Shipman]

Gelatine comes from collagen in animal remains, usually those of cattle. It is a transparent substance without taste or smell. It dissolves in hot water and then forms a jelly when cool. It is used in many products and processes. Agar from seaweed is frequently used for bacterial cultures today.

Harley Street: A well known thoroughfare in central London recognized as the location of many doctors' and medical specialists' consulting rooms.

King's College Hospital, London: It started as a Medical Department of King's College in 1831. The hospital moved to a new building in South East London in 1913.

NOTES 2C

Charing Cross Hospital, London: Founded in 1818 in Suffolk Street as the West London Infirmary and Dispensary by Benjamin Golding. In 1823 it moved to Villiers Street near The Strand, and in 1827 was renamed the Charing Cross Hospital. It has since moved to Fulham.

Quinine was the usual medicine of the time for malaria, both as a preventative and for its treatment. Quinine itself was extracted from Cinchona bark.

Quack nostrums: These were medical remedies, usually having little or no effect on the symptoms for which they were sold.

Castor oil: Two drops of castor oil, frequently used as a purgative, would have little adverse effect on a person, whereas **strychnine**, in large doses would cause violent convulsions and almost certainly death. Castor oil comes from the castor oil plant (*Ricinus communis*). Strychnine, which has a bitter taste, is an alkaloid and comes from the tree *Strychnos Nux-vomica* and from similar plants. Strychnine was often used as a rat and pest poison.

Tetanus can cause lockjaw and affects the nervous system. It is an acute infection caused by *Clostridium tetani* bacteria which gain access to the body through a wound or bite. Spasm, rigidity and stiffness of the muscles are symptoms, with the jaw and neck first affected. Convulsions and breathing difficulties can follow.

Hippocratic smile and risus sardonicus (literally a sardonic smile): Tetanus can cause an exaggerated grin as a result of involuntary facial muscle contraction.

The Andaman Islands are situated in the Bay of Bengal.

Blair Island is South Andaman of the Andaman Islands. Port Blair is the capital.

NOTES 2D

Toxicologist: An expert on poisons and their antidotes.

Arrow poison: This may have been extracted from the arrow poison frog *Phyllobates horribulus* or some other similar species of poisonous frog. Death will occur if the poison, a batrachotoxin, is ingested. The toxins can cause convulsions, hallucinations, heart failure and death.

Curare: A poison which is extracted from the bark of trees of the species *Strychnos toxifera* and *Chondodendron*. It is an alkaloid drug, which if absorbed through the skin causes breathing difficulties. Death can rapidly follow.

Queen of England: The answer is Eleanor of Castille, Queen of Edward the First.

Prussic acid, which is another name for hydrocyanic acid, is a deadly and fast acting poison. It dissolves readily in water to give a clear colourless liquid and has the smell of bitter almonds. It occurs naturally in some seeds and stones of fruits, such as those of the peach and apricot.

Devil's-foot Root: Fictional.

A pharmacopoeia describes drugs and their use and effects.

Ordeal poison: In a trial by ordeal, a person accused of a crime or misdemeanour was sometimes given poison to prove their innocence or guilt. Survival was supposed to prove innocence; suffering and probably death indicated guilt.

Ubanghi country: The Ubanghi is a river which flows through the Congo region of Africa before joining the River Congo.

A chip box is one made of small pieces of thin wood.

NOTES 2E

Cocaine is an addictive drug which can be injected in solution, sniffed or smoked in the form of crack. It is an alkaloid which comes from the plant *Erythroxylon coca* and other species, and the chemical name is methyl-benzoylecgonine. When taken it creates a feeling of euphoria. It is used in medicine as an anaesthetic.

Opium comes from the opium poppy *Papaver somniferum*. Morphine is derived from it and used in medicine as a pain killer. Opium was popularized in Britain from about 1680 by the physician Thomas Sydenham. It became widely available as laudanum in Victorian times. Illegal opium dens operated at one time in an area occupied by the Chinese around Limehouse Causeway and Pennyfields near the docks in East London.

De Quincey's book *Confessions of an English Opium-Eater* was published in 1822, although the text first appeared in the *London Magazine* in 1821.

Laudanum is a tincture of opium.

Waxed or waned: Increased and decreased.

Bang or bhang (hashish): This is now better known as cannabis, and it comes from the Indian hemp plant. It is usually smoked for its relaxing properties but can be harmful if its use is continued.

NOTES 2F

Morphia (synonymous with morphine): An alkaloid medical painkiller derived from opium. It can be addictive.

Carafe: This was a glass container usually holding drinking water. The application of cold water to the skin would have been a shock but at least it would have had the effect of diluting the acid. Warm water would have been better, especially some containing a dissolved alkali to neutralize the acid.

Oil: In the first incident (re Ian Murdoch), a soothing oil, such as olive oil, on the skin would have given only minor relief. In the second incident (re Baron Gruner) the oil used may have been olive oil, or carron oil (a mixture of lime water, linseed oil and eucalyptus oil).

The **Specialist** was probably an expert on burns or acid burns.

Cooling medicine: This may have been a concoction to cool the blood, a solution to provide a tranquillising effect, or perhaps just a simple soothing or cooling drink, such as barley water and cream of tartar.

Incense can contain natural gums and spices, or the aromatic leaves and branches of plants which exude pleasant and highly scented fumes when heated or burnt.

Prague: At the time the capital of Czecho-Slovakia in Bohemia.

The langur is an Indian monkey. It has a slender body and long tail. It is also known as the Entellus Monkey, or Hanuman revered by Hindus.

Anthropoid means 'similar to man'.

Typhoid fever: See Enteric fever under Notes 2A.

Consumption or Tuberculosis: An infectious lung disease, due to the bacterium *Mycobacterium tuberculosis*, though other parts of the body can be affected. Poor nutrition and/or bad housing exacerbate the spread of the disease. It was common in Holmes' day.

NOTES 2G

Clasp knife: A folding pocket knife.

Stitches: Catgut, silk, horsehair and wire have been used at various times for surgical stitching.

Linen compress: A linen pad used as a bandage to stop the wound bleeding.

An aortic aneurysm is a swelling in the wall of the aorta and is liable to burst.

Nosebleed and ruddy complexion: Nosebleeds can be an indication of disease, but they also occur in persons who have too many red blood corpuscles.

Diabetes: There are various types of diabetes, and a sufferer can have various symptoms which may include drowsiness, thirst and weight loss. With *diabetes mellitus* there is a lack of insulin production by the pancreas, which causes a surfeit of sugar in the blood and urine. The importance of insulin was discovered by Banting, Best and Macleod in 1921-22.

Monomania is a mental condition whereby the sufferer becomes obsessed by a particular subject dominating his or her mind.

Nervous lesions: Alteration or damage to tissues and organs or their function, are known as lesions.

Catalepsy: The sufferer adopts an unnatural posture and becomes silent and still as if in a trance.

Nitrite of amyl (amyl nitrite): A sweet smelling yellow liquid. For the treatment of *Angina pectoris*, the vapour is inhaled so as to rapidly reduce the blood pressure.

Rigidity of muscles: In other words rigor mortis. This stiffening begins within a few hours of death. After it has become established, it gradually disappears, the whole process taking at least twenty-four hours. Variations in temperature and other factors can alter the timescale.

"WE STROLLED ABOUT TOGETHER."

The Resident Patient

3 THE OLD, THE PRECIOUS AND THE BURIED

ARCHAEOLOGY, GEOLOGY AND ANTHROPOLOGY

ARCHAEOLOGY

Archaeology is the study of the lives and customs of peoples of the past, including their human remains, artefacts and buildings. Examination and interpretation of items excavated from historic sites has enabled archaeologists to discover more about the culture and everyday life of our ancestors. Most of the references to archaeology in the Sherlock Holmes stories and novels are confined to The Hound of the Baskervilles, though there are also some in two other adventures (The Three Garridebs, The Devil's Foot).

The first, an exciting story about the mysterious 'spectral' hound, takes place in the beautiful rural county of Devon, and reference is made to the neolithic and early settlements on Dartmoor, where the tale unfolds. *The whole steep slope was covered with grey circular rings of stone, a score of them at least.* Stapleton, a schoolmaster and collector of natural history specimens, explains to Watson that they are the remains of homes once inhabited by Neolithic man [3A]. *"These are his wigwams with the roofs off. You can even see his hearth and his couch if you have the curiosity to go inside."* He then goes on to tell Watson that this early man *"learned to dig for tin* [3A] *when the bronze* [3A] *sword began to supersede the stone axe".*

Later, when writing to Holmes in London, Watson describes in graphic detail the appearance of the Moor and its archaeological remains. *On all sides of you as you walk are the houses of these forgotten folk, with their graves and the huge monoliths* [3A] *which are supposed to mark their temples.* He then goes on to say that *if you were to see a skin-clad, hairy man crawl out from the low door, fitting a flint-tipped* [3A] *arrow on to the string of his bow, you would feel that his presence there was more natural than your own.*

Watson also mentions an open, grassy space *on which rose two great stones, worn and sharpened at the upper end, until they looked like the huge, corroding fangs of some monstrous beast.* Further in this long report to Holmes, Watson tells him that Doctor Mortimer *has been excavating a barrow* [3A] *at Long Down, and has got a prehistoric skull* [3A] *which fills him with great joy.*

Later still we learn something of the eccentric and legal-minded Mr. Frankland, who spends a fortune on litigation, it seems for the fun of it. He takes a dim view of Dr. Mortimer's excavations on Long Down, and it is rumoured, intends to prosecute him *for opening a grave without the consent of the next of kin.* In relating this to Holmes, Watson sees the comical aspect of the situation.

In The Three Garridebs when Holmes and Watson arrive at Nathan Garrideb's residence, they are taken into a room that serves as Nathan's home museum: *Here was a case of ancient coins. There was a cabinet of flint instruments. Behind his central table was a large cupboard of fossil bones. Above was a line of plaster skulls with such names as "Neanderthal* [3A]*," "Heidelberg* [3A]*," "Cro-Magnon* [3A]*" printed beneath them.* Nathan is polishing a coin with a piece of chamois leather [3B]: *"Syracusan* [3B] *– of the best period,"* he explained holding it up.

Another mystery (The Devil's Foot) has a West country setting, but this time in Cornwall. *In every direction there were traces of some vanished race which had passed utterly away, and left as its sole record strange monuments of stone, irregular mounds* [3A] *which contained the burned ashes of the dead, and curious earthworks which hinted at prehistoric strife* [3B]. At one stage in this investigation, Holmes decides to put the case aside for a time and as he informs Watson, *"devote the rest of our morning to the pursuit of Neolithic man* [3A]*".* He then proceeds for two hours to discuss *celts* [3B], *arrow heads* [3B]*, and shards* [3B]*, as lightly as if no sinister mystery was waiting for his solution.* One can only sympathize with Watson.

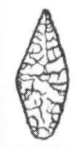

Arrowheads

GEOLOGICAL LANDSCAPE

The second part of A Study in Scarlet opens with a description of an arid and repulsive desert. *From the Sierra Nevada* [3B] *to Nebraska* [3B]*, and from the Yellowstone River* [3B] *in the north to the Colorado* [3B] *upon the south, is a region of desolation and silence.* The account tells us that this barren area is grim and is a land of despair. *Snow-capped and lofty mountains, dark and gloomy valleys, swift-flowing rivers, jagged canyons, and enormous plains* [3C] *which in winter are white with snow, and in summer are grey with the saline alkaline dust* [3C] make it a miserable and inhospitable area.

Dartmoor [3C] features in two of the stories (Silver Blaze, The Hound of the Baskervilles). The second of these has the following description: *The road in front of us grew bleaker and wilder over huge russet and olive slopes, sprinkled with giant boulders.* The moor has treacherous bogs, and the danger of such a bog is shown in the adventure when Cyril Stapleton disappears. As Holmes and Watson follow his trail, we have this account: *a false step plunged us more than once thigh-deep into the dark, quivering mire, which shook for yards in soft undulations around our feet* (The Hound of the Baskervilles).

Lower Gill Moor [3C] is a similar type of country, though in the North of England; and there is also moorland around Poldhu Bay [3C] in Cornwall. *It was a country of rolling moors, lonely and dun-coloured* [3C] (The Devil's Foot). In the same story there is a description of the view from Holmes' and Watson's holiday cottage, which overlooks Mounts Bay [3C] *with its fringe of black cliffs and surge-swept reefs.*

Other geological references are recorded in the stories. For example it is mentioned that the more modern wings of Baskerville Hall are built of black granite [3D]; and the extraction and processing of Fuller's earth [3C] is given as an explanation by Colonel Lysander Stark for the use of the hydraulic press [3D] in The Engineer's Thumb. A clay bearing stream passes Birlstone Manor House, with the flow of water sweeping liquid clay further downstream (The Valley of Fear); and processed clay is present in an ancient Chinese eggshell porcelain [3D] saucer taken by Doctor Watson to Baron Gruner, an expert collector of such items (The Illustrious Client). As Holmes hands it to Watson, he tells him *"This is the real eggshell pottery of the Ming dynasty* [3D]. *No finer piece passed through Christie's* [3D]."

THE NOBLE METALS

Gold is a precious metal which has been mined and used to make highly prized artefacts from earliest times. It is also the one which the early alchemists tried to create from baser metals, in their search for the elusive 'philosopher's stone' [3D].

Silver is also a valuable metal, and articles made of both this and gold have always been coveted by the rich and honest. Criminals though have also often been attracted to them, usually for monetary gain, but sometimes to obtain something precious and beautiful which they can keep but do not have to pay for.

In A Case of Identity Holmes proffers Watson snuff from a bejewelled box of old gold [3D]. Watson comments on the splendour of the box, and Holmes explains that *"it is a little souvenir from the King of Bohemia* [3D] *in return for my assistance in the case of the Irene Adler papers"*. In the same story Miss Mary Sutherland wears small, round, hanging gold ear-rings, and her newspaper advertisement requesting the whereabouts of Hosmer Angel, mentions that he wears a gold Albert chain [3D].

Godfrey Norton has a gold watch in A Scandal in Bohemia, and Holmes is given a sovereign of gold [3D] by Irene Adler, when he acts as a witness at her marriage to Godfrey Norton. Previously in this story, half sovereigns have been promised to various cab and coach drivers for the journey to the church of St. Monica.

A Study in Scarlet has several references to gold. A woman's gold wedding ring is found at the scene of Enoch Drebber's murder. Also in his pockets are a gold watch by Barraud [3D]; a gold Albert chain [3D]; a gold ring with a masonic device – that of the lodge to which Drebber belonged; and a gold pin with a bulldog's head on it. Further in the story Holmes indiscreetly bribes an off-duty policeman with a golden half-sovereign.

The Sign of Four has several references to the treasure owned by a wealthy Indian rajah, who has a store of gold and silver which he keeps in the vaults of his palace in a northern province of India. We are told that *"He is of a low nature and hoards his gold rather than spends it."* The rajah decides to hedge his bets, and divides his treasure into two in case one half of it is captured. One part of the collection, which contains many jewels and a gold coronet, he sends to Agra [3D]. In the same adventure when Jonathan Small is offered a chance to share in this great treasure, his thoughts turn to the instant wealth that he would gain. *"How my folk would stare when they saw their ne'er-do-weel coming back with his pockets full of gold moidores* [3D]*."*

BASE METALS

A criminal gang of coiners has been using a combination of metals to produce thousands of counterfeit half-crowns [3E] in The Engineer's Thumb, as revealed by the large masses of nickel and tin discovered stored in an outhouse. Another counterfeiting reference occurs in Shoscombe Old Place when Holmes refers to a case in which he *"ran down that coiner by the zinc and copper filings in the seam of his cuff"*.

Dartmoor is an area where tin mines were once common and Mrs. Stapleton tells Holmes of *" an old tin mine on an island in the heart of the mire"* where her husband may be hiding out (The Hound of the Baskervilles). In another instance *a huge driving-wheel and a shaft half-filled with rubbish showed the position of an abandoned mine.*

PRECIOUS STONES AND PEARLS

Precious stones have been created by geological action over millenia. They have been sought-after from early times, and items of jewellery and objects encrusted with them have been manufactured for years, as they are still made today.

Pearls of course have a different origin. They come from sea molluscs, and originate from a simple grain of sand, parasitic larva or other irritant inside the creature. The pearls are formed layer by layer as a deposit of calcium carbonate is built up and polished within the shell of the oyster or mussel.

Pearls feature largely in The Sign of Four. Miss Mary Morstan consults Holmes over a mystery puzzling her. Regularly each year she has been sent by post *"a very large and lustrous* [3E] *pearl without any clue as to the sender"*. Later in the story we learn that before his death, Major Sholto asked his sons to recompense Captain Morstan's daughter (that is Mary) whom he felt had been cheated from her share of the Agra treasure. He refers to a chaplet [3E] tipped with pearls as a suitable item for recompense. When the sons find the treasure hidden in Sholto's house, Thaddeus honours his father's wishes by sending Miss Morstan *"a detached pearl at fixed intervals so that at least she might never feel destitute"*. Towards the end of the case, Jonathan Small explains how the treasure's owner, a wealthy rajah, had placed *"the most precious stones and the choicest pearls in an iron box and sent it by a trusty servant to the fort at Agra"*. When Small and his partners in crime eventually get their hands on this fabulous treasure, he remarks that *"it was blinding to look upon them"*, a reference to jewels consisting of diamonds [3E] of the first water [3E], including *"the Great Mogul, said to be the second largest stone in existence,"* and very fine emeralds, rubies, carbuncles, sapphires, agates, beryls, onyxes, cats'-eyes, turquoises and other stones [3E,3F]. There were also some very fine pearls, *"twelve of which were set in a gold coronet* [3F]*."* However there seems a little inconsistency here, in that a chaplet (though this can be used as a term for a garland worn on the head) tipped with pearls has been referred to earlier, whereas now it appears to be a gold coronet embellished with pearls (The Sign of Four). Holmes has a gold snuff box with a great amethyst ³ᶠ in the centre of the lid, and also a ring with a remarkable brilliant ³ᶠ which sparkles upon his finger (A Case of Identity). He also owns a neat little leather case containing burglars' instruments. One of these is a diamond-tipped glass-cutter ³ᶠ (Charles Augustus Milverton).

The story line of The Mazarin Stone ³ᶠ revolves around the yellow diamond of this name which Lord Cantlemere has had stolen. Needless to say Holmes conducts another successful investigation, and at the end cannot resist playing a small joke on the rather stuffy Lord Cantlemere by placing and 'discovering' the stone in his Lordship's coat pocket.

Another adventure with a precious stone at the centre of the case, is The Blue Carbuncle ³ᶠ. The jewel is stolen, and a goose intended for Christmas swallows it but is confused with another: the ensuing mix-up leading Holmes to track down the thief.

A third story where valuable stones ³ᶠ are featured, is the case of The Beryl Coronet. The coronet ³ᶠ has been 'lost' by Alexander Holder, a banker in whose safe-keeping (as security for a loan) it had been placed. He is beside himself with anxiety and asks Holmes to investigate the loss. The coronet itself has thirty-nine enormous beryls ³ᶠ and fine gold chasing ³ᴳ.

Early in another adventure (A Scandal in Bohemia), we learn that the King of Bohemia ³ᴰ wears a cloak secured at the neck with a brooch which consists of a single flaming ³ᴳ beryl. The King offers Holmes an emerald snake ring at the conclusion of the case, but Holmes declines it, preferring to have a photograph of Irene Adler instead. A gold-pin with a bulldog's head and with rubies as eyes is found amongst the contents of Enoch Drebber's pocket (A Study in Scarlet).

ANTHROPOLOGY

Anthropology embraces the study and science of the development of mankind, including the evolutionary period.

The case of The Cardboard Box gives Holmes the opportunity to mention his two short monographs on the variation and differences of the human ear. *"Each ear is as a rule quite distinctive, and differs from all other ones."* He says that these monographs [3G] appeared in the Anthropological Journal [3G] of the previous year.

A further case (The Disappearance of Lady Frances Carfax) has Watson confused and irritated when Holmes sends him a telegram asking for *a description of Dr. Shlessinger's left ear. "Holmes' ideas of humour are strange and occasionally offensive, so I took no notice of his ill-timed jest."* But it is a serious enquiry and Holmes has to find another source for the information he requires.

In The Sign of Four, Holmes finds an entry in a gazetteer describing the inhabitants of the Andaman Islands [3G] in the Bay of Bengal. *The aborigines of the Andaman Islands may perhaps claim the distinction of being the smallest race upon this earth, though some anthropologists prefer the Bushmen of Africa* [3G]*, the Diggers of America* [3G]*, and the Terra del Fuegians* [3G].

In another adventure Mortimer tells Holmes and Watson (The Hound of the Baskervilles) that he and Sir Charles Baskerville discussed *"the comparative anatomy of the Bushman and the Hottentot* [3G]*"*.

CRANIOLOGY

Craniology is the study of the bones and structure of the head. The Hound of the Baskervilles has a few references to this subject, which is a favourite source of study for the enthusiastic Doctor Mortimer. At his first meeting with Holmes at Baker Street, he tells him with a directness which is amusingly stated: *"You interest me very much, Mr. Holmes. I had hardly expected so dolichocephalic a skull* [3H] *or such well-marked supra-orbital development* [3H]*. Would you have any objection to my running my finger along your parietal fissure* [3H]*? A cast of your skull* [3H]*, sir, until the original is available, would be an ornament to any anthropological*

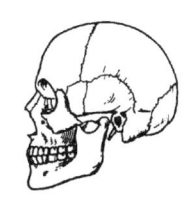

museum. It is not my intention to be fulsome, but I confess that I covet your skull."

A little later Doctor Mortimer congratulates Holmes on his observation concerning the use of *The Times* newspaper for supplying the lettering in a message sent to Sir Henry Baskerville. To demonstrate that a detailed knowledge of a subject can make an individual an expert, Holmes questions Mortimer: *"you could tell the skull of a negro from that of an Eskimo?"* Mortimer answers that he could: *"Because that is my special hobby. The differences are obvious. The supra-orbital crest* [3H]*, the facial angle* [3H]*, the maxillary curve* [3H]*, the - - -."* And so Holmes proves his point.

NOTES 3A

Neolithic man (New Stone Age man): Lived in the period immediately before the use of metals became established (circa 2,300 BC). Neolithic men made flint and stone implements for hunting and domestic use, and carried out animal husbandry and crop cultivation to provide food for themselves.

Tin: A silvery white metal. It was mined on Dartmoor from about 1150.

Bronze: An alloy consisting of copper and tin.

The monoliths (sometimes called menhirs) are giant standing stones on the Moor. There are also many stone circles.

Flint-tipped: Flint was used by Stone Age man for arrow and axe heads. The mineral was readily available, very hard and could be 'worked' or knapped to the required shape and sharpness. It was a time-consuming task requiring some skill, so to be caught knapping on a long Dartmoor evening was certainly no disgrace.

Barrows or Mounds: The graves of these early men were known as barrows. Many can still be seen today.

Prehistoric skull: In Britain, one from the period before about 43AD.

Neanderthal, Heidelberg, Cro-Magnon: These names refer to the remains of early prehistoric men found in various parts of Europe. All of them were of especial interest to archaeologists and anthropologists and have provided clues in helping to explain the evolution of man.
- **Neanderthal:** Human skeletal remains were found in the Rhine Valley in 1856.
- **Heidelberg:** The lower jaw of an early man was found in 1907 near the German tower of Mauer.
- **Cro-Magnon:** Several skeletons were found in 1868 at Les Eyzies-de-Tayac in France.

NOTES 3B

Chamois leather: An especially soft leather from the skin of the Chamois, a variety of antelope. It is ideal for polishing objects.

Syracusan coin: Syracuse in Sicily was founded as a Greek colony in 734 BC.

Earthworks for prehistoric strife: This is probably a reference to earthworks (hillforts) made as a fortification for defence.

Celts: In this context, probably a reference to prehistoric implements made of flint, bone, stone or other hard materials. Used as tools, they were fashioned with chisel-shaped edges.

Arrowheads were obviously made by early man for hunting purposes, but they could also be used against enemies. Flint, and later, metal was used for these heads.

Shards are the broken fragments of pots and utensils used as containers by neolithic man.

Sierra Nevada: A mountain range in California, USA.

Nebraska: A central state of the USA.

Yellowstone River, USA: It is over 600 miles in length, running from Wyoming to just inside North Dakota, where it joins the Missouri.

Colorado River, USA: Over 1000 miles long from the Rocky Mountains, through the Grand Canyon (Arizona) and into Mexico.

NOTES 3C

Salt plain: This is probably a reference to the Great Basin. Evaporation of water can leave behind deposits of chemical salts and common salt (sodium chloride).

Saline alkaline dust: Dust containing salt and compounds, such as soda and potash, that can neutralize acids.

Dartmoor is a large moorland area in Devon of over 300 square miles. Features are grey jagged granite tors, treacherous bogs or mires, pastures grazed by ponies and white-faced Dartmoor sheep; and here and there the relics of mining activity and the scattered remains of prehistoric settlements.

Lower Gill Moor: Probably fictitious.

Poldhu Bay: To be strictly accurate it should be Poldhu Cove.

Dun-coloured is a grey/brown colour.

Mounts Bay: This bay stretches from Gwennap Head to Lizard Head at the toe of Cornwall.

Fuller's earth is a clay of some fineness, with good absorbent properties. It was used for fulling cloth, hence the name, and in a wide range of other industrial processes. There are two types of Fuller's earth, the one mentioned here is almost certainly the variety that contains montmorillonite.

NOTES 3D

A hydraulic press operates by using water pressure to create and magnify the force, which is then utilized in the pressing operation. One of the earliest successful hydraulic presses was invented by the engineer, Joseph Bramah, who obtained a patent for it in 1795.

Granite: A rough, hard grey igneous rock mineral deposit composed mainly of feldspar, quartz and mica.

Eggshell porcelain: A very thin translucent porcelain.

Ming dynasty: Chinese dynasty from 1368 to 1644.

Christie's: An auction house founded in London in 1766 by James Christie.

Philosopher's stone: In alchemy, the mythical substance required to convert base metals to gold or silver.

Old gold is a description of the metal when tarnished and lacking the brilliant golden colour of new gold.

An Albert chain is a watch chain named after Prince Albert, Queen Victoria's consort.

Bohemia: A province of Austria-Hungary until 1918, and then part of the republic of Czechoslovakia.

Sovereign: A British gold coin worth twenty shillings, i.e. £1.

Barraud watch: Several companies with the name of Barraud were working in London during the 19th century.

Moidores are Portuguese gold coins. In the eighteenth century they could be used in England as currency.

Agra: Indian city in Uttar Pradesh. Best known as the location of the Taj Mahal.

NOTES 3E

Half-crown = 2s. 6d. = 12½ pence in present currency. In Holmes' time, genuine half-crowns would have been made of real silver. Later (from 1947) they were made from cupro-nickel.

Lustrous: The magnificence or iridescence of the pearl.

A chaplet is a necklace or garland.

Diamonds are highly prized precious stones of extreme hardness and composed of pure carbon. They sparkle when cut by a skilled craftsman.

First water: Diamonds of the best quality and greatest purity.

Emeralds can be translucent or transparent and receive their green colouration from chromium traces. They are a form of beryl, containing beryllium and aluminium silicate.
Rubies are extremely hard (though less hard than a diamond), and are red in colour and made of corundum, which is a mineral consisting mainly of aluminium oxide (alumina).
Carbuncles are dark red in colour, or of a reddish hue; are soft in texture and are usually cut with concave surfaces. They are a variety of almandines or precious garnets. Their composition includes alumina, silica and iron protoxide.
Sapphires, like rubies, are composed of corundum, an alumina bearing deposit. Many are blue, but they can be other colours except red, which is the colour of the ruby.

NOTES 3F

Agates consist of a type of silica or chalcedony. There is quite a wide variation of colours and they polish easily to bring out their colour. Like some diamonds, they are used for industrial purposes as well as for jewellery manufacture.
Beryls contain beryllium and aluminium silicate. They tend to be greenish in colour, though there are varieties in other hues. Some of the stones are opaque and known as 'common beryls' and the transparent ones as 'precious beryls'.
Onyxes, like agates, are also composed of a type of silica. They are usually black and white or brown and white in alternating layers or bands.

Cats' eyes: These are stones, which when cut so as to have a dome-shaped top, show a similarity to the eyes of a cat, due to an optical effect known as chatoyancy. The stones mentioned in this adventure are probably chrysoberyls (or cymophanes), which consist mainly of beryllium aluminate, and are green or greenish yellow in colour. Cats' eyes can also be produced from quartz.

Turquoises are the colour of their name – blue to bluish-green. The mineral contains aluminium phosphate, with copper from which their colouration comes. They can be opaque or slightly translucent.

An **amethyst** consists of quartz (SiO_2) and its colour can be purple, blue or shades of the two.

A **brilliant** is a diamond cut to emit maximum brilliance.

Diamond-tipped glass-cutter. The hardness of a diamond permits it to cut glass.

The Mazarin stone (also called the Crown diamond) was probably named after Cardinal Jules Mazarin, a 17th century papal diplomat and French statesman.

The **carbuncle** in this story is extraordinary in that it has a blue colouration. See note under 3E.

Coronet: An ornamental band usually of metal, or a small crown.

NOTES 3G

The chasing is the embossed design of the gold on the coronet.

Flaming beryl: A gem of intense brightness.

Monographs are published works concerned with a single subject.

The Anthropological Journal: A publication with this title has not been traced, so is probably fictitious.

The Andaman Islands are now controlled by India.

Diggers (Indians): These were mainly members of the Paiute tribe, who occupied parts of California and other states. Their main source of food came from roots dug up from the ground, hence the name.

Terra del Fuegians: Inhabitants of an island off the coast of South America. Tierra del Fuegia means 'Land of Fire'. 'Terra' in the novel is an anglicized version of Tierra.

Bushmen and Hottentots: These were nomadic/semi-nomadic groups of people native to South Africa and with a similar style of language. Comparison of their physical and cultural characteristics were of great interest to anthropologists of the time as they searched for evidence of the evolutionary development of mankind.

NOTES 3H

Someone with a **dolichocephalic skull** is a long-headed person.

A supra-orbital development is the structural arrangement above the eye socket.

The parietal fissure is the narrow groove between the two parietal bones in the cranium.

Cast of the skull: one method of making an accurate reproduction of the skull of a living person, would have been to model the head in clay and take a cast from this. The procedure would have required some artistic skill, and sketches with accurate measurements of the head with calipers would have been essential. A cast, perhaps using a two-piece mould could then have been made using plaster of Paris.

The supra-orbital crest is a ridge above the eye cavity.

The facial angle: A specific craniometric measurement based on the relative spacing of the nasion, gnathion and the Frankfurt horizontal plane.

The maxillary curve is the line of the jaw bone.

4 LADY NICOTINE AND THE THREE PIPE PROBLEM

TOBACCO AND SMOKING

There are numerous references to smoking and tobacco in the Sherlock Holmes stories. At the time that they were written, the majority of Victorian and Edwardian gentlemen liked to relax and enjoy a smoke in their leisure time. Together with a glass of port or some other alcoholic beverage in the hand, a feeling of affability and of being at peace with the world could be attained.

At that time the harmful effects of smoking were generally unknown, and cigars, cigarettes and pipes were readily available and their use accepted without question. Sherlock Holmes and Watson were no exception in their appreciation of a good smoke, and they frequently enjoyed tobacco in all the above forms, and also used snuff in the course of their adventures. Several major and minor characters in the stories, we are told, also often indulged in the habit. From both personal use and professional interest, Holmes was an expert on individual tobaccos and their ashes, and wrote a comprehensive monograph on the subject, of which more later.

Smoking has very early origins. Herodotus (ca 484-425 BC) reported that the Scythians inhaled the smoke of hemp [4A]; and Pliny (23 - 79) said that the smoke from coltsfoot, *Tussilago farfara*, had a beneficial effect on coughs [4A]. Even today the plant is sometimes called Coughwort and is still used as a popular cough remedy, often in conjunction with other plant extracts. The preserved Egyptian mummy of Ramesses II was found to contain chopped up tobacco leaves, although there has been no evidence to show that the plant was smoked by the Ancient Egyptians.

The Mayas, an early civilisation in Central America, used pipes to blow tobacco smoke [4A] towards the sun, a procedure probably followed as a form of worship. In the fifteenth century, Indians smoking tobacco were discovered in Cuba by Christopher Columbus, and the use of tobacco gradually spread to Europe, brought in by the Spanish.

An unusual use of tobacco, as it would seem to us, was to adopt it as an unofficial form of currency for barter [4A] between buyers and sellers in many parts of the world. In Britain, though, it was used in the conventional way.

It is likely that an ordinary sailor may have been the first to bring the tobacco habit to Britain in 1556, although Sir Walter Raleigh (1552-1618) is often regarded as the one responsible for encouraging its popular use here, especially by introducing the clay [4A] pipe as a receptacle in which to smoke it. As the tobacco became more widely available and cheaper, the size of the pipe bowl increased in size.

The tobacco plant *Herba Nicotiana* was named after the French ambassador Jean Nicot (1530-1600), who popularized smoking in France. The word 'nicotine' is derived from Nicot's name, but the actual nicotine, a harmful alkaloid, was not purified and isolated from the plant until 1828.

Today there are two important varieties of the tobacco plant. These are *Nicotiana rustica*, from which the 'oriental tobaccos' are derived, and *Nicotiana tabacum*, which is the most widely grown species of the plant, and from which the majority of smoking products are now manufactured.

These products are made from the dried and cured leaves of the plant. They are rolled, shredded and made into cigarettes and cigars, or packed for use in pipes, self-made cigarettes or for chewing. Snuff requires further processing. The tobacco in all these products contains many chemicals, most of them harmful, especially the nicotine which is present and causes addiction, and carbon monoxide which is released when it is smoked. There is a serious risk of cancer and heart disease from smoking, and tobacco also contains damaging mutagens [4A] which can cause genetic disorders in the body. The higher the tar content of the tobacco, the more harmful it is. It is interesting to note that the residual ash after tobacco was burnt sometimes provided a helpful clue in an investigation by Sherlock Holmes.

TOBACCOS

Various types of tobacco are mentioned in the stories. On Holmes' and Watson's first meeting (A Study in Scarlet), they discuss sharing lodgings together, and Holmes asks Watson initially: *"You don't mind the smell of strong tobacco, I hope?"* When Watson tells him that he always smokes 'ship's' [4A] himself their future compatibility is assured. Then in Black Peter, the same variety, *"half an ounce of strong ship's tobacco"* is found inside a sealskin tobacco-pouch which Inspector Hopkins discovers at the scene of the crime.

In another episode (A Scandal in Bohemia), Holmes, in disguise, helps the ostlers with their horses at Irene Adler's stables, and for his efforts is rewarded with *two fills of shag tobacco* [4A]. Holmes obviously has a liking for this tobacco, because in a later story (The Hound of the Baskervilles), he asks Watson to order for him from Bradley's [4A], one pound [4B] *"of the strongest shag tobacco."*

Late one night Holmes calls on Watson and immediately identifies the pipe tobacco that Watson has been smoking, as Arcadia. This he identifies from the fluffy ash down the front of Watson's coat (The Crooked Man).

Thaddeus Sholto smokes an Eastern tobacco with a balsamic odour [4B], and asks Holmes, Watson and Miss Morstan *"I trust that you have no objection to tobacco smoke"* (The Sign of Four).

Holmes tells Watson that the Grosvenor mixture used by Grant Munro for his pipe, costs eightpence [4B] an ounce [4B], and so he deduces that Munro *"has no need to practise economy."* (The Yellow Face)

CIGARS

The widespread smoking of cigars in Britain developed from about the time of the Napoleonic war. Many cigars were made by hand in Victorian times, and some knowledge was required to blend the various tobaccos to produce a cigar that was both choice and provided a satisfying smoke. Basically each cigar consisted of three parts: namely the wrapper (a hollow cylinder of rolled leaves) and a filler and binder inside.

Although we know that Holmes and Watson liked to smoke a cigar occasionally, we are not told if they always smoked the best brands. Still, whatever type they smoked, here are some examples of those occasions when they did enjoy a cigar. The following two excerpts come from The Empty House.

Having just solved his case, Holmes addresses his colleague. *"And now, Watson, I think that half an hour in my study over a cigar may afford you some profitable amusement."* Then back in the old familiar Baker Street study, Watson describes the scene as Holmes consults his index of biographies: *He turned over the pages lazily, leaning back in his chair and blowing great clouds of smoke from his cigar.*

" I smoked a cigar and waited behind a tree" (The Boscombe Valley Mystery)

In The Boscombe Valley mystery, Holmes tells Watson to *"Light a cigar, and let me expound"*. Then in the same story the detective explains: *"I found the ash of a cigar, which my special knowledge of tobacco ashes enabled me to pronounce as an Indian cigar. Having found the ash, I then looked round and discovered the stump among the moss where he had tossed it. It was an Indian cigar, of the variety which are rolled in Rotterdam* [4B].*"* Watson then queries the likely use of a cigar holder[4B] by the smoker, to which Holmes replies: *"I could see that the end had not been in his mouth. Therefore he used a holder. The tip had been cut off, not bitten off, but the cut was not a clean one, so I deduced a blunt penknife."*

A Trichinopoly [4B] cigar features in A Study in Scarlet when Holmes informs Gregson and Lestrade that the murderer smoked a cigar of this type, leaving a black, flaky ash, which Holmes, with his expert knowledge, is easily able to identify.

Abel White, in India, smokes cheroots [4B] on his veranda (The Sign of Four), and in the same adventure, we learn that Major Sholto also smokes cheroots.

Holmes deduces that there is a ventilator in a room at the residence of Doctor Roylott before he even sees the room, because he has learnt beforehand that Julia Stoner when in her own room adjoining Doctor Roylott's, could smell his cigar (The Speckled Band).

Some differences between cigars are noted by Holmes in The Resident Patient. In Blessington's (the murder victim's) room, Inspector Lanner has found four cigar ends in the fireplace. Holmes looks at these and also inside Blessington's cigar case. This contains a single cigar, which Holmes identifies as a Havana [4B]. The ones from the fireplace, he observes, are *"of the peculiar sort imported by the Dutch from their East Indian colonies"*. He examines the four cigar ends closely with his lens, proclaiming that two were smoked from a cigar holder and two without a holder. Two of them have been cut with a slightly blunt knife, *"and two have had their ends bitten off by a set of excellent teeth,"* he states. From all this Holmes concludes that a cold-blooded murder has been committed, and not a suicide!

Holmes is exhilarated when he hears that Doctor Mortimer has exercised typical Holmesian deductive skills, and inferred that Sir Charles Baskerville must have stood in one place for five or ten minutes *"because the ash had twice dropped from his cigar"* (The Hound of the Baskervilles). Previously in the same story, we learn that Sir Charles took a regular nocturnal walk, smoking a cigar.

Cigars are also mentioned in The Sign of Four and The Valley of Fear. In the former, Watson offers Detective Athelney Jones a cigar; and in the latter, Mr. Douglas, who has been in hiding in his own house, is keen to smoke when at last he emerges from his hiding place. He says *"you'll guess what it is to be sitting for two days with tobacco in your pocket and afraid that the smell will give you away"*. Holmes then kindly hands him a cigar to satisfy his craving.

Another man, Captain Croker, is under stress because he knows that Holmes has discovered that he killed the cruel Lord Brackenstall. Holmes with some sympathy and the knowledge that smoking can soothe the nerves, instructs Watson to *"give him a cigar. Bite on that, Captain Croker, and don't let your nerves run away with you. I should not sit here smoking with you if I thought that you were a common criminal, you may be sure of that."* (The Abbey Grange)

Finally it should be mentioned that an eccentricity of Holmes is to keep his cigars in the coal-scuttle [4B], and his tobacco in the toe end of a Persian slipper [4B] (The Musgrave Ritual) .

PIPES

Over the centuries pipes have been made of many materials. Clay, wood and stone (some of them intricately carved), and also porcelain have all been popular with smokers in various lands. Clay pipes only became common in Britain in the sixteenth century with the introduction of tobacco. Those with long stems became known as churchwarden pipes [4B] in the late 17th and early 18th centuries because some church officials (churchwardens) liked to smoke them after their meetings. Eventually when the ubiquitous clay pipe began to fall from favour, the briar and meerschaum started to increase in popularity.

In illustrations, Holmes is sometimes depicted smoking a meerschaum [4C] pipe, but this is not actually mentioned in any of the stories or novels about him. Instead we are told that he smokes a clay pipe – an 'old and oily' one in A Case of Identity, and indeed this type of pipe is frequently used by him. However it appears that the pipe he favoured depended on his mood at the time, so in The Copper Beeches Watson sees that he is using his *long cherry-wood pipe which was wont to replace his clay when he was in a disputatious rather than a meditative mood.*

A black clay pipe also features in The Hound of the Baskervilles. Watson returns to Baker Street one evening to find the room filled with smoke, and as he succinctly puts it: *My first impression as I opened the door was that a fire had broken out.* The cause of course is Holmes, who, by smoking his clay pipe, is creating *acrid fumes of strong, coarse tobacco.*

On one occasion when Doctor Watson is at home, having just enjoyed a late night pipe, Holmes calls on him unexpectedly. After requesting a bed for the night and declining a meal, Holmes says to him, *"I'll smoke a pipe with you with pleasure"* (The Crooked Man). Although Watson does not always appear to be quite such an enthusiast for tobacco as Holmes, in A Study in Scarlet he is again found smoking his pipe.

It is not unusual for smokers to dislike the smoking habits of others around them, and at times, Watson is no exception. His account in The Valley of Fear reports that *Sherlock Holmes lit the unsavoury pipe which was the companion of his deepest meditations.*

In The Yellow Face Grant Munro leaves his pipe in Holmes' consulting rooms. *"A nice old briar* [4C] *with a good long stem of what the tobacconists call amber* [4C]. *I wonder how many real amber mouthpieces there are in London. Some people think a fly* [4C] *in it is a sign."* Holmes examines it closely and observes that it has been *"twice mended: once in the wooden stem and once in the amber. Each of these mends, done, as you observe, with silver bands, must have cost more than the pipe did originally."* From this Holmes deduces that Munro prizes it highly. He remarks that perhaps only watches and bootlaces (!) have more individuality than pipes. After further perusal of the pipe he then concludes that Munro must be *"a muscular man, left-handed , with an excellent set of teeth, careless in his habits, and with no need to practise economy"*. His explanation of these observations is entertaining and should be read in the actual story for a full understanding.

Holmes often uses his pipe-smoking to aid his thinking. Watson asks him what he is going to do to solve the problem in **The Red-Headed League**, and Holmes replies: *"To smoke. It is quite a three-pipe problem, and I beg that you won't speak to me for fifty minutes."* A really good simile is found in the same story. *There he sat with his eyes closed and his black clay pipe thrusting out like the bill of some strange bird.*

Other characters in the adventures smoke pipes as well. In **The Sign of Four** Jonathan Small has just taken out his pipe, when he is jumped on by two Sikh soldiers; and in **Silver Blaze**, we learn that John Straker smokes an A.D.P. pipe [4C].

Thaddeus Sholto favours a hookah pipe [4C] in **The Sign of Four**. The smoke produced from it is bubbled through rose-water [4C] contained in the bowl of the pipe: the aroma probably adding to the scented atmosphere of the room. In the same story, a briar-root [4C] pipe is also mentioned.

CIGARETTES

Cigarettes date from at least 1575, when they were smoked by Mexicans who rolled their tobacco in paper. Later they were introduced into Spain, from where they spread to other parts of Europe. They arrived in Britain about the time of the Crimean War and gradually grew in popularity, until by the time of the First World War, they had become the most common type of tobacco product used in the country. Women started smoking cigarettes about 1900, with female indulgence in the habit, often using cigarette holders, becoming more common during the First World War.

Cigarettes and their ash are essential clues in the solving of the adventure of The Golden Pince-Nez; while in The Boscombe Valley mystery Holmes has *a caseful of cigarettes which need smoking*; and in another investigation he lights a cigarette before throwing himself down into an armchair. (A Scandal in Bohemia)

Doctor Mortimer uses self-rolled cigarettes in The Hound of the Baskervilles, and on seeing this Holmes cannot resist telling him:
"I observe from your forefinger [4C] *that you make your own cigarettes. Have no hesitation in lighting one."* Mortimer then proceeds to do so.

In the same adventure, Holmes knows Watson has found his retreat on Dartmoor when he finds a cigarette stub beside the path leading to the hut he is using. He tells Watson: *"If you seriously desire to deceive me you must change your tobacconist; for when I see the stub of a cigarette marked Bradley* [4A]*, Oxford Street, I know that my friend Watson is in the neighbourhood."* Bradley is Watson's (and Holmes') usual tobacconist in London, but of course the shop must have had other customers, so Holmes could never have been absolutely certain that a 'Bradley' cigarette end originated from Watson.

A Scandal in Bohemia has a splendid sentence expressing Holmes' delight at finding the origin of a watermark on a note he has received. *His eyes sparkled, and he sent up a great blue triumphant cloud from his cigarette.*

THE HOLMES MONOGRAPH ON TOBACCO ASH

Holmes informs us that he is an expert in identifying tobacco products and the ash residues that they produce. We learn in The Boscombe Valley Mystery that he has written a learned monograph on the subject. This covers the ashes of 140 different varieties of pipe, cigar, and cigarette tobacco. Its title as given in The Sign of Four is: *"Upon the distinction between the Ashes of the various tobaccos",* and we are told here that this authorative work contains coloured plates illustrating the various differences in the ashes. *"If you can say definitely, for example, that some murder had been done by a man smoking an Indian lunkah* [4C], *it obviously narrows your field of search. To the trained eye there is as much difference between the black ash of a Trichinopoly* [4B] *and the white fluff of bird's-eye* [4C] *as there is between a cabbage and a potato."*

The way in which Holmes analysed his ash samples is not described, but he must certainly have observed the texture, smell, and amount of ash produced, as well as making use of chemical analysis to determine the constituents of the residue. If he found a stub, as in The Boscombe Valley Mystery, this was a bonus as a clue to identity. Also the ring [4D] from a cigar would enable him to identify one very quickly.

SNUFF

The habit of snuff-taking in Europe started in Spain and came to Britain via France in the 17th century. Snuff, which is usually inhaled nasally (though there are other ways to use it), is basically chopped and powdered tobacco, often flavoured with oils, spices, and perfumes, and sometimes with added medicinal compounds.

At first, snuff was made by the individual consumer himself. He had to buy ordinary tobacco, often in the form of a rope of intertwined leaves. A length was then chopped off and ground into powder by hand, using an instrument (grater) called a snuff rasp. Later though, snuff was milled and ground mechanically at the manufacturers, and then supplied to local tobacconists ready for sale. The advantage of this mass-produced snuff was its finer texture compared with the coarser home-made variety.

The adventures have a few references to snuff. In The Red-Headed League Holmes notices that Jabez Wilson takes it, and in the conversation that follows, Wilson partakes of a huge pinch of snuff in order to refresh his memory. Mycroft, the brother of Holmes also takes snuff (The Greek Interpreter); and in A Case of Identity Holmes proffers a gold snuff box [4D] to Watson, saying: *"Take a pinch of snuff, Doctor."*

SMOKING APPAREL AND SMOKING CONCERTS

Culverton Smith (The Dying Detective) is wearing a piece of headgear known as a 'smoking cap' when Doctor Watson unexpectedly calls on him. In his account Watson describes the appearance of the man: *A high bald head had a small velvet smoking-cap poised coquettishly upon one side of its pink curve.*

In Victorian times, it was not acceptable for women to smoke; and men were frequently left to smoke in the dining room after a meal, or banished to another room where they could partake of their habit in a male only environment. Often the men would change into special 'smoking clothes', consisting of a jacket, perhaps of velvet, with wide lapels and richly embroidered decoration. Also they sometimes put on caps similar to the one described in the extract above.

In The Valley of Fear reference is made to smoking concerts, which John Douglas would go to in the village of Birlstone. These concerts, usually attended by men, were musical events where smoking was permitted. It is unlikely though that cigarettes were smoked by the instrumentalists between pieces, least of all the woodwind and brass players, as they needed quite a different kind of puff to give a satisfactory musical performance.

NOTES 4A

Hemp: There are several types of hemp plant, but this reference is probably to the cannabis or marijuana plant *Cannabis sativa*.

Health benefits: Early on, tobacco was regarded as a medicinal plant to cure ailments and disease of all types. At the time of the 1665 plague, reputedly the boys at Eton College were compelled to smoke every morning to fight the disease. If they did not do so, they were whipped.

Tobacco smoke was often recommended to revive people rescued from drowning, and into the early nineteenth century some army doctors carried a special apparatus for this purpose. A publication of circa 1819, under the heading of *Dr. Hawses' Method of Restoring to Life drowned Persons*, suggests that *Tobacco-smoke be thrown gently up the fundament, with a proper instrument or the bowl of a pipe, covered so as to defend the mouth of the assistant.*

Barter: In the 17th and 18th centuries, in the American states of Virginia and Maryland, tobacco was officially recognized as a currency.

Clay pipes: An article describing the method of making clay pipes can be found in *The Countryman* (February/March 1997, pp. 63-7).

Mutagens: Substances that can cause mutation in cells.

Ship's tobacco was produced by a number of companies that often sold it under the name of Navy Cut. It was a strong coarse tobacco, which got its name from the roll of tobacco that sailors often liked to purchase and cut into slices. Player's Navy Cut and Navy Mixture became a great favourite of smokers, and John Player and Sons, the manufacturer, sold both in packaging bearing the familiar design of the head of a bearded seaman 'framed' in a lifebelt. Other manufacturers included Nicholls of Chester and Richard Lloyd and Sons of London. W.D. and H.O. Wills of Bristol produced a sweetened variety which they sold under the name of Windlass.

Shag tobacco was a shredded dark coarse tobacco used in pipes. It was made from a heavy large type of leaf, which produced a strong smelling aroma when smoked.

Bradley's was a tobacconist in Oxford Street – a fact we learn in The Hound of the Baskervilles.

NOTES 4B

One pound in weight = approximately 454 grams.

A balsamic odour is one with a fragrant oleo-resin smell.

Eightpence (8d.) = about three pence (3p.) in present currency.

One ounce = approximately 28 grams.

Rotterdam: A port and city in the Netherlands.

Cigar holders were also sometimes made of meerschaum, which is a mineral containing magnesium silicate. It is light and porous and can be highly polished.

A Trichonopoly cigar is one named after the Indian city of Trichinopoly (now Tiruchirappalli) in Madras. The area is renowned for its cigars and cheroots manufactured from the locally grown dark tobacco.

A cheroot is a slender tubular cigar with both ends cut like a cigarette.

Havana cigars are made in Cuba, and are invariably regarded as the finest that can be bought. The Vuelta Abajo supplies some of the best tobacco for these.

A coal-scuttle is a sturdy household container, usually of metal, and was used to hold the fuel to keep an open fire supplied and burning. It was also utilized to carry coal from the storage area, cellar or bunker to the fireplace.
Containers for cigars were usually more conventional than Holmes' coal-scuttle. Boxes to hold a number of cigars were generally made of wood, often Gabon wood or an aromatic cedar. From about 1890, single cigars were sold in glass tubes, or contained in a wooden box with a sliding lid. Novelty cigar containers could also be found.

A Persian slipper is a soft shoe which can easily be slipped on and off, and has a curled end at the toe.

Churchwarden pipes: These long elegant pipes are still made today by craftsmen using metal moulds, and Devon clay as the material from which the pipes are formed.

NOTES 4C

Meerschaum pipe: Usually only the bowl is made of meerschaum – a mineral composed of hydrous magnesium silicate (sepiolite). The material (white, yellow or grey) can be carved and polished. When the pipe is smoked the colour tends to change to a darker hue.

Briar (or brier) pipes are made from the root of the Tree Heath, *Erica arborea* which is grown in the Mediterranean region of Europe and other parts of the world. The wood is suitable for making pipes as it is very hard, has a fine texture which can be highly polished, and chars only slowly.

Amber is a hard fossilized resin the colour of golden honey and contains succinic acid. The remains of ancient creatures or plants are sometimes found preserved in amber.

Fly in amber. The amber has probably come from a conifer tree now extinct. A fly could easily become trapped and preserved in the sticky substance which was exuded and solidified so long ago. This fly can therefore be many thousands of years old and so indicate a true amber pipe mouthpiece of some distinction.

A hookah pipe has an oriental origin. It consists of a vase containing water which is sometimes scented. A tube and the bowl of the pipe are attached to this vase, and the smoke passes through the scented water before reaching the mouth of the smoker by means of a long flexible tube also attached to the vase. It has been recorded that Inderwick's of London, a tobacconist founded in 1797, once made a hookah pipe with seven mouthpieces, so that the various members of a family could smoke it at the same time.

Rose-water is a perfume produced from rose petals.

An A.D.P. pipe is one manufactured by Alfred Dunhill, a renowned tobacconist of London.

Forefinger nicotine stain: The familiar yellow mark often seen on the finger of a dedicated smoker.

A lunkah is a strong-flavoured Indian cheroot.

Bird's-eye: A type of tobacco which has the fibre and leaf-ribs cut together. Suitable for both pipes and cigarettes, it was sold in tins by the large London company of Salmon and Gluckstein.

NOTES 4D

The cigar ring or band. A well-printed or embossed label encircling the cigar and bearing the trade mark and/or name of the maker. It was once considered a breech of smoking etiquette to smoke a cigar with the band in place, thus displaying the brand name to public gaze.

Snuff boxes have been made of a variety of materials, such as wood, pewter, precious metals and ivory. Some are ornately decorated or carved, and others are sometimes encrusted with jewels. The owner's name or initials may also be engraved onto the box.

The Greek Interpreter

5 BULLETS, SHARP BLADES AND BLUNT INSTRUMENTS

BALLISTICS AND WEAPONRY

Murder is the major feature of many of the Sherlock Holmes stories, and the methods and weapons used to carry out the crimes are varied and occasionally unusual. Injury, both minor and severe, is also common throughout the adventures, and the weapons that are employed to inflict wounds on victims, but do not cause death, are often the same as the ones that are used to kill.

"In his hand he held a pistol"

BALLISTICS AND DETECTION

The technical term 'ballistics' in criminal investigations is applied to the scientific study of the way in which guns are fired, and the way that bullets or projectiles behave in motion and the consequences on reaching their target.

When a crime using firearms has been committed, the ensuing

investigation will usually require certain questions to be asked Some of these questions may be as follows: What type and make of gun, rifle or other weapon was used? What was the calibre and type of bullet, and size of shot or pellets? Were any spent cartridge cases, splinters or shrapnel found at the scene? Are there any special marks on any bullets, if some were found at the scene of the incident, or extracted from the victim? If so, can these be compared with any suspect weapon in order to obtain a match? From what distance did the assailant fire his or her weapon, and what was the trajectory of the bullets or projectiles? What was the type of wound caused to the victim, and from what angle did any of the bullets enter the body? Are there any powder burns visible, and can analysis find any traces of powder on the clothing of the victim or of any suspect who is apprehended? From the answers to these questions, it may be possible to deduce the location of the gunman at the time of the shooting and an understanding of how exactly the crime was committed.

Holmes' method of examining a case where a firearm has been fired, is well illustrated in **The Dancing Men**. After Mr. Hilton Cubitt is found shot dead and his wife discovered with serious injuries, Holmes has the task of finding out who fired the gun: *for the revolver lay upon the floor midway between them.* At the time of the crime, the room itself and the passage outside was full of smoke and the smell of powder. *"Unless the powder from a badly fitting cartridge happens to spurt backwards, one may fire many shots without leaving a sign,"* says Holmes, after seeing that the bullet has been fired at Cubitt from the front, and that there is no powder-marking of his dressing gown or hands. His wife has stains on her face but none on her hand, he notes. In addition Holmes notices that an extra bullet, which cannot be accounted for, has been fired and gone right through the lower part of the window frame. This observation proves the presence of a third person – a point of some importance as Mrs. Cubitt is under suspicion of committing her husband's murder.

FIREARMS USED BY HOLMES AND WATSON

Watson, who served in the army, (as did Conan Doyle during the Boer War) is frequently asked by Holmes to bring his army revolver with him or to have it ready to hand. All the cases where Watson needs his gun are too numerous to list, but some examples are given below:

In one story (The Sign of Four) Holmes asks Watson: *"You have not a pistol, have you?"* To which Watson replies *"I have my stick"*. Holmes thinks the stick may be useful if they trace and trap the killers in their hiding place, but on this occasion, he himself intends to rely on a more convincing weapon: *He took out his revolver as he spoke, and, having loaded two of the chambers, he put it back into the right-hand pocket of his jacket.* Later on Holmes asks: *"Have you a pistol, Watson?"* *"I have my old service-revolver in my desk"* comes the reply.

Holmes himself only occasionally carries a revolver, otherwise he favours his loaded hunting-crop [5A] which was his favourite weapon (The Six Napoleons).

Running up, I blew its brains out. (The Copper Beeches)

In another adventure (The Problem of Thor Bridge) Holmes asks Watson: *"Have you your revolver on you?"* *I produced it from my hip-pocket, a short, handy, but very serviceable little weapon.* Holmes requires the revolver for an experiment, and carries out the following test with it. Using a large stone as a weight and a string tied to the gun, he suspends the weight over the bridge parapet. *He raised the pistol to his head, and then let go his grip. In an instant it had been whisked away by the weight of the stone, had struck with a sharp crack against the parapet, and had vanished over the side into the water.* The simple ballistics aspect of this case is shown in the following extract, and leads to the arrest of the governess. *"A revolver with one discharged chamber and a calibre which corresponded with the bullet was found on the floor of her wardrobe,"* states Holmes, in explaining the evidence against her. However his own investigation of the events in the case leads him to come to an entirely different conclusion.

The Speckled Band has a reference to a specified ammunition when Holmes asks Watson to bring his revolver. *"An Eley's No. 2* [5A] *is an excellent argument with gentlemen who can twist steel pokers* [5A] *into knots."* The 'gentleman' in question here being the unpleasant Doctor Grimesby Roylott who has just demonstrated his anger and strength by bending Holmes' poker, which the detective manages to re-straighten.

An amusing incident occurs in The Sign of Four. In the cab on the way to Thaddeus Sholto's house, Watson tells Mary Morstan anecdotes of his Afghanistan adventures, but in a slightly garbled version it seems. In his written account of the case he records that: *To this day she declares that I told her one moving anecdote as to how a musket looked into my tent at the dead of night, and how I fired a double-barrelled tiger cub at it.*

At least on one occasion Sherlock Holmes fires off cartridges in an irresponsible display of eccentric behaviour, and vandalizes the Baker Street consulting room: *Holmes, in one of his queer humours, would sit in an armchair with his hair-trigger* [5A] *and a hundred Boxer* [5A] *cartridges, and proceed to adorn the opposite wall with a patriotic V.R.* [5A] *done in bullet-pocks.* Of this, Watson quite naturally and thoroughly disapproved (The Musgrave Ritual).

OTHER FIREARMS AND AMMUNITION

Neil Gibson has a formidable array of firearms of various shapes and sizes which he has accumulated in the course of an adventurous life (The Problem of Thor Bridge); and an old-fashioned pin-fire revolver [5A] is found amongst various items left behind in Wisteria Lodge by tenants who have done a moonlight flit (Wisteria Lodge).

Other firearms are also mentioned in The Sign of Four. These are the musket [5A] and the carbine [5A]. Jonathan Small recounts how, when on guard duty in India (Agra) [5A], he laid down his musket and was instantly leapt on by two Sikhs, who were his fellow guards. *"One of them snatched my firelock* [5A] *up and levelled it at my head."* Later on Small throws his weapon at a merchant fleeing for his life: *"I cast my firelock between his legs as he raced past, and he rolled twice over like a shot rabbit."* One of the guards at the convict camp on the Andaman Islands where Small was held, carried a carbine on his shoulder, and raised it as Small attacked him.

In **The Reigate Puzzle** Holmes demonstrates the importance of ballistics in his summing up of the case: *"The wound upon the dead man was caused by a shot from a revolver fired at a distance of something over four yards. There was no powder-blackening* [5A] *on the clothes."* From these observations Holmes concludes that *"Alec Cunningham had lied when he said that the two men were struggling when the shot was fired".*

In **The "Gloria Scott"** a multiplicity of weapons is used in the mutiny of the prisoners on board the ship. The arms are distributed to the rebels by a sham chaplain as James Armitage (old Trevor) explains in a letter to his son: *By the third day we had each stowed away at the foot of our bed a file, brace of pistols, a pound of powder, and twenty slugs.* The soldiers guarding the convicts on board had muskets, we learn, but two of them are shot while trying to fix their bayonets.

Several types of ammunition are mentioned in the adventures. For example in **The Sign of Four** on finding some of Tonga's poisoned darts, Holmes remarks that *"I would sooner face a Martini bullet* [5B]*, myself".* Also in the same story a Jezail bullet [5B] is given as the cause of Watson's war wound in the leg. Yet **A Study in Scarlet** informs us that he was *struck on the shoulder by a Jezail bullet.* Did he therefore have two war wounds or was Conan Doyle inconsistent in his descriptions?

But it is another type of bullet that is used to murder the Honourable Ronald Adair, whose *head had been horribly mutilated by an expanding revolver bullet* [5B] (**The Empty House**). As the adventure proceeds Holmes discovers a powerful air-gun [5B], with which Colonel Moran intended to kill him. On examining the mechanism, he remarks: *"An admirable and unique weapon,"* said he, *"noiseless and of tremendous power. I commend it very specially to your attention, Lestrade, and also* *the bullets which fit it."*

Godfrey Emsworth tells his army comrade, James Dodd, that at Buffelspruit [5B] *"I got an elephant bullet [5B] through my shoulder"* (The Blanched Soldier).

In The Valley of Fear John Douglas appears to have been murdered by the discharge of a shotgun, as at the scene of the crime, *lying across his chest was a curious weapon, a shotgun with the barrel sawed off a foot in front of the triggers. Fired at close range, and the triggers had been wired together so as to make the simultaneous discharge more destructive.* Holmes quickly identifies the American manufacturer of the double-barrelled shotgun, which fires buckshot cartridges [5B], as the Pennsylvania Small Arms Company [5B].

EXPLOSIVES

The positioning of highly explosive dynamite [5B] was the intended method of blowing up the house of New York fruit importer Castalotte, when he stands up to the threats of The Red Circle, a blackmailing Italian secret society.

Blasting powder [5B], used in mines and quarries, is referred to in one of the adventures (The Valley of Fear). The powder is placed and fired by the Scowrers [5C] at the house of Chester Wilcox and his family, but they are away from home and escape the explosion.

In The "Gloria Scott" there is a scene during the mutiny when the first mate threatens to blow up the ship with everyone on board if he is harmed. The convicts find him with *a match-box in his hand seated beside an open powder-barrel, which was one of a hundred carried on board.*

CUDGELS AND STICKS

Cudgels [5C], bludgeons [5C], sticks and life preservers all make their appearance on numerous occasions in the stories. When Watson is in Montpellier [5C] on behalf of Holmes, he gets into a difficult situation and is attacked, but guess who comes to the rescue. *My senses were nearly gone before an unshaven French ouvrier [5C] in a blue blouse, darted out from a cabaret [5C] opposite, with a cudgel in his hand, and struck my assailant a sharp crack over the forearm, which made him leave go his hold* (The Disappearance of Lady Frances Carfax). The 'French ouvrier' turns out to be none other than Holmes in disguise.

In one story Holmes is called to investigate what appears to be a vicious night-time murder and robbery at the home of the Brackenstalls. Lady Brackenstall tells Holmes that Sir Eustace, her husband, *"evidently heard some suspicious sounds, and he came prepared for such a scene as he found. He was dressed in his shirt and trousers, with his favourite blackthorn [5C] cudgel in his hand"* (The Abbey Grange).

Another knight, Sir George Burnwell, it seems was always prepared, for he kept a life-preserver [5C] on the wall of one of his rooms, as any security-conscious householder certainly should! On one occasion he intends to use it too, but Holmes has come well-prepared and *claps a pistol to his head before he can strike* (The Beryl Coronet).

Holmes calls Oberstein a 'cunning dog', but he is also a murderer, as the detective discovers when Colonel Walter is persuaded to provide him with a witness statement: *Oberstein had a short life-preserver. As West forced his way into the house Oberstein struck him on the head. The blow was a fatal one* (The Bruce-Partington Plans).

A stick or cudgel under a different name can be found in another of the adventures when the unsavoury Mr. Latimer draws *a most formidable-looking bludgeon loaded with lead from his pocket* (The Greek Interpreter). He then waves it about in front of Mr. Melas, the translator, in a menacing way.

Walking sticks also feature as common weapons in the stories. Jim Browner described how he used an oak stick to attack his unfaithful wife, Mary and her lover Alec Fairbairn. The pair were in a boat, and although Fairbain *"jabbed at me with an oar, I got one in with my stick, that crushed his head like an egg. I struck again, and she lay stretched beside him"* (The Cardboard Box).

The body of Bartholomew Sholto is discovered with an unusual weapon nearby. *By his hand upon the table there lay a peculiar instrument – a brown close-grained stick, with a stone head like a hammer, rudely lashed on with coarse twine* (The Sign of Four).

When Count Sylvus attempts to use his stick for assault (The Mazarin Stone), he is thwarted by a wily Holmes, who has cleverly positioned a wax dummy [5D] in his likeness in the window of his Baker Street room. The Count approaches the dummy with the intention of attacking it, thinking it is Holmes, but he is surprised by the real detective addressing him from the doorway. *"Don't break it, Count! Don't break it!" The assassin staggered back, amazement in his convulsed face. For an instant he half raised his loaded* [5D] *cane once more, as if he would turn his violence from the effigy to the original.* A ferocious attack on Holmes by two men with sticks is reported in an evening newspaper. But Holmes, who is an excellent fencer, uses his single-stick [5D] to good advantage in defending himself (The Illustrious Client). Afterwards, in hospital he tells Watson *"I'm a bit of a single-stick expert, as you know. I took most of them on my guard. It was the second man that was too much for me."*

POKERS

A handy weapon of the time was the poker, which was widely available because coal fires were so ubiquitous. Consequently it appears several times in the adventures, and although it is not always used as a weapon, it does cause the murder of a night watchman in The Stockbroker's Clerk. *The man's skull had been shattered by a blow from a poker, delivered from behind.*

Sir Eustace Brackenstall suffers a similar fate. *"His head was knocked in with his own poker",* Inspector Hopkins tells Holmes when he arrives at the scene of the crime (The Abbey Grange).

In yet another story (The Sign of Four) this instrument is applied in a non-aggressive way by Doctor Watson as he uses it to open a box retrieved from the river, expecting it to contain the Agra treasure. But he is in for a surprise when the lid is lifted.

KNIVES

The knives that can be found in the canon come in various types, shapes and sizes, though they are not always used as weapons. *"When you catch this fellow, you will find that he has one of these multiplex knives* [5D] *in his possession",* Holmes observes when investigating a case in which a corkscrew has been used to remove a wine bottle cork, hence giving rise to the deduction above (The Abbey Grange).

He had ripped up my trousers with his pocket-knife. This event occurs when Watson is injured and Holmes acts quickly to discover the seriousness of his wound (The Three Garridebs).

A sealing-wax knife is used for murder in The Golden Pince-Nez. *The instrument with which the injury had been inflicted lay upon the carpet beside him. It was one of those small sealing-wax knives* [5D] *to be found on old-fashioned writing-tables, with an ivory handle and a stiff blade.*

In The Six Napoleons a special type of knife appears to be the murder weapon used on Pietro Venucci. This was described as *a horn-handled* [5D] *clasp-knife* [5D]*, lying in a pool of blood beside him.* However when the murderer is caught and searched, a different kind of knife is found on him: *A search of his clothing revealed nothing save a few shillings* [5D] *and a long sheath knife* [5D]*, the handle of which bore copious traces of recent blood.*

Another knife is found in the bundle retrieved by Cecil Barker in The Valley of Fear. Holmes undoes the bundle and removes its contents one by one until *he laid upon the table a long, deadly, sheathed knife.*

The murder of Eduardo Lucas is carried out using a curved Indian dagger in The Second Stain, but even more exotic and curious weapons crop up in another adventure (The Sign of Four). Details of these can be found below.

OTHER WEAPONS

An unusual type of killer is successfully dealt with when Holmes and Stackhurst push a boulder into a pool in order to exterminate a threatening specimen of the Lion's Mane or *Cyanea* [5D] lurking there. *One flapping edge of yellow membrane showed that our victim was beneath it* (The Lion's Mane).

A boulder is also used as a weapon in A Study in Scarlet. *As Drebber passed under a cliff a great boulder crashed down on him, and he only escaped a terrible death by throwing himself upon his face.* Although it is not plainly stated in the account, it is implied that Jefferson Hope caused the boulder to fall.

A meat cleaver [5D] is a most unpleasant weapon, and this or a similar implement is wielded to remove Victor Hatherley's thumb as he tries to escape, although a more vicious attack is intended. *"I saw the lean figure of Colonel Lysander Stark rushing forward with a lantern in one hand, and a weapon like a butcher's cleaver in the other. I was hanging with my fingers in the window slot and my hands across the sill, when his blow fell"* (The Engineer's Thumb).

The corridor walls of Hurlstone are decorated with old weapons. When Reginald Musgrave suspected there was an intruder in the library, he tells Holmes: *"I picked a battleaxe [5D], and then, leaving my candle behind me, I crept on tiptoe down the passage and peeped in at the open door"* (The Musgrave Ritual).

Lethal and dangerous weapons in The Sign of Four include a blowpipe that fires a poisoned dart that ends the life of Bartholomew Sholto; while stone-headed clubs and poisoned arrows are mentioned in the encyclopedia entry on the Andaman islanders that Holmes reads out to Watson. Tonga, who is one of these islanders and a faithful follower of Jonathan Small, carries a long bamboo spear, which Small utilizes in combination with some coconut matting [5D] to create a sail. Before this though, Small has used his wooden leg as an extraordinary cudgel with which to kill one of the guards on the Island.

Another unusual weapon, a harpoon [5D], is used to murder Black Peter. *It had been snatched down from a rack on the wall. Two others remained there, and there was a vacant place for the third* (Black Peter).

A metal spanner that is heavy and easy to grip makes an ideal weapon for attack or defence. However Menzies, the unfortunate Crow Hill Mine engineer, is shot before he can retaliate with the spanner he holds. Bravely, having just seen the Scowrers shoot the mine manager, he rushes with an iron spanner at the murderers, but meets the same unhappy end (The Valley of Fear).

Cheeseman's is a house in Sussex which has *a fine collection of South American utensils and weapons.* It transpires that the weapons are of some significance in the case that Holmes is called to investigate. He later admits that *"when I saw that little empty quiver* [5D] *beside the small bird-bow* [5D], *it was just what I expected to see".* It turns out that curare, *"or some other devilish drug",* as Holmes puts it, placed on the tip of an arrow could have threatened the life of the young child involved in this mystery which is really based on jealousy (The Sussex Vampire).

Two unusual weapons are used in another case (The Abbey Grange). Theresa, the Australian nurse, has a decanter thrown at her by Sir Eustace Brackenstall, and she informs Holmes that Lady Brackenstall *"never told me of those marks on her arm that you saw this morning, but I know very well that they come from a stab with a hat-pin* [5D] *."*

In The Final Problem Holmes describes to Watson how his life was threatened three times. First of all by a two-horse van driven deliberately at him by a hit-and-run driver; and secondly by the occurrence of a suspicious incident: *"As I walked down Vere Street a brick came down from the roof of one of the houses and was shattered at my feet".* He also tells Watson about the third attempt: *"Now I have come round to you, and on my way I was attacked by a rough with a bludgeon* [5C] *."* On this occasion Holmes retaliates and knocks down the ruffian, who is quickly arrested.

NOTES 5A

A hunting-crop is a short horse-whip with a curved handle and leather loop. The handle can be weighted with metal.

Eley's No. 2: This is probably a reference to the .442 cartridge which was used in the Webley, Tranter and Enfield revolvers.

Poker: A long manufactured metal rod with a handle, used for stoking or disturbing the coals of an open fire to encourage heat. In this case it is made of steel, but brass was also a popular metal for such an instrument.

A hair-trigger is one that is so sensitive that it only needs a slight squeeze to activate the mechanism to fire the gun.

Boxer cartridges are named after Colonel Edward Boxer (of Woolwich Arsenal) who designed them. They were often used in the Snider Rifle.

V.R. is the Royal cipher (Victoria Regina) of Queen Victoria, the reigning monarch during the period of the early cases. If Queen Victoria had known about Holmes' target practice, she may well have appreciated his patriotism, but like Watson, would not have been amused by the bullet-pocked wall.

The pin-fire revolver had a mechanism that drove a pin into the fulminate used as a detonator in the cartridge.

A musket is a long-barrelled firing weapon. It can be either muzzle or breech-loaded, and can be activated according to the type of gun. Examples are the matchlock (lighted with a match applied to powder); the flintlock (using a flint to create a spark); or the percussion musket. A wheel-lock version was also in use at one time.

A carbine was often used by cavalry troops. It is a type of short rifle.

Agra is a city in India on the River Jumna.

A firelock is a musket in which a flintlock and steel is used to create a spark to ignite the powder.

Powder-blackening: Caused by burnt explosive powder leaving traces on the clothes of the victim by the force of the shot.

NOTES 5B

A Martini bullet: Frederich Martini was a 19th century civil engineer and inventor, whose breech-loading mechanism was fitted to the Henry rifle to produce the Martini-Henry rifle.

A Jezail bullet was one used in the Afghan musket of that name.

Expanding revolver bullet or soft-nosed bullet, sometimes known as a Dum-

Dum bullet, causes particularly severe injuries as it disintegrates and spreads on entering the body. The bullet was invented by Captain Bertie Clay at the ammunition factory at Dum-Dum in India.

The air-gun uses compressed air to fire the bullets. The idea of such a weapon probably originated from the blowpipe, and a version using bellows was invented in the 16th century. The spring-operated air gun by G.W.B. Gedney is probably the type mentioned in this Sherlock Holmes story, and was patented in 1892.

Buffelspruit: A location in the Aliwal North Eastern Cape Province area of South Africa.

Elephant bullet: When large game such as elephants were hunted, the ammunition used required an extra hardness. For this purpose early ball-shaped bullets were coated in pewter.

Buckshot cartridges: These are cartridges containing lead shot and are normally used for shooting deer.

The Pennsylvania Small Arms Company: Untraced, probably fictitious.

Dynamite: A high explosive formulated by the Swedish chemist Alfred Nobel and patented in 1867. At this time it contained nitro-glycerine and kieselguhr (a form of silica) and a small amount of an alkaline solid such as chalk (calcium carbonate) or magnesium carbonate.

Blasting powder: This is what is termed a low explosive, as only simple ignition is required, without the need of a detonator. Gunpowder and Faversham powder are examples.

NOTES 5C

The Scowrers: The Eminent (or Ancient) Order of Freemen is a fraternal secret society. Fictitious, but probably based on 'The Molly Maguires'. This was a real Irish/American secret society for Pennsylvania miners in operation before 1880.

Cudgel: A short heavy thick stick.

Bludgeon: A short thick heavy stick.

Montpellier: The capital of the Hérault department in Southern France.

French ouvrier: A workman.

Cabaret: Probably a tavern in this context.

Blackthorn is a thorny shrub with small white flowers. Its fruit is the blue-black sloe berry.

Life-preserver: A weighted stick.

NOTES 5D

Wax dummy: Craftsmen at waxworks, such as Madame Tussauds, used wax moulding techniques to make true-to-life figures.

Loaded cane: A cane, perhaps of bamboo and therefore hollow, allowing it to be internally weighted with lead or another metal.

Single-stick: A stick which could be used to fence with.

Multiplex knife: A multipurpose knife with various implements attached. For example it would have been useful in Victorian/Edwardian London to have had an instrument to remove stones from a horse's hoof.

Sealing wax knife: One that could be used to spread the hot wax when sealing a letter or package, and also perhaps to break and lift the seal on any that were received.

Horn: The bony outgrowth from the heads of cattle, sheep and other mammals. There was a plentiful supply of this material, which could be carved or shaped for knife handles before man-made materials such as plastic became common.

Clasp-knife: A folding pocket-knife.

A **Shilling** (1/-) is 5 pence in present currency.

Sheath knife: A sharp bladed knife kept in a case or cover often made of leather.

***Cyanea capillata* or the Lion's Mane** is a stinging jellyfish which can sometimes be found off the British coast. In the past it was rare but today numbers are increasing slightly. The jellyfish derives its name from its tentacles, which resemble a lion's mane and have been known to reach over forty feet long on an exceptional specimen. (See also Chapter 11)

Meat cleaver: A short-handled broad-bladed chopper used by butchers for cutting up meat.

Battleaxe: A short heavy hand-held axe used as a weapon.

Coconut matting: A coarse fibre floor covering made from the thick layer of hair (coir) obtained from coconuts.

Harpoons: Barbed weapons like spears with attached ropes and floats, were used in the whaling industry. At first they were thrown by hand, but later, guns were used to fire them.

Quiver: A container to hold an archer's arrows.

Bird-bow: This was probably a bow for shooting bird-bolts (blunt-headed arrows) at birds.

Hat-pin: A long, large-headed pin used by women to keep a hat in place on the head.

6 THE ACID TEST

CHEMISTRY

Chemistry is Holmes' favourite science and the one in which he really excels. From the early days of his career, the chemistry laboratory is the place where he feels very much at home, and as mentioned at the beginning of this book, is the location where Doctor Watson first meets him. Saint Bartholomew's Hospital [6A] in London is the scene of this introduction, which leads to the long-standing friendship of the two men and the many intriguing adventures that they pursue together. Before this historic meeting however, Watson has already been told by Stamford, a former colleague, that Holmes is a first-class chemist (A Study in Scarlet). On arrival at the hospital, Stamford and Watson pass through the entrance and approach the chemistry laboratory together, and Watson later records what they saw on entering the room: *This was a lofty chamber, lined and littered with countless bottles. Broad, low tables were scattered about, which bristled with retorts* [6A], *test-tubes, and little Bunsen lamps* [6A], *with their blue flickering flames.*

The above passage gives us some idea of what the inside of the laboratory looked like and helps us to imagine it, but tells us nothing of the smells and sounds that would have been present there as well. What would these have been? In truth we do not really know, but it is likely that visitors on entering the laboratory may have been able to recognize one or more of its highly distinctive emanating odours. With a sniff of the air, the pungent smell of hydrochloric acid, or the choking

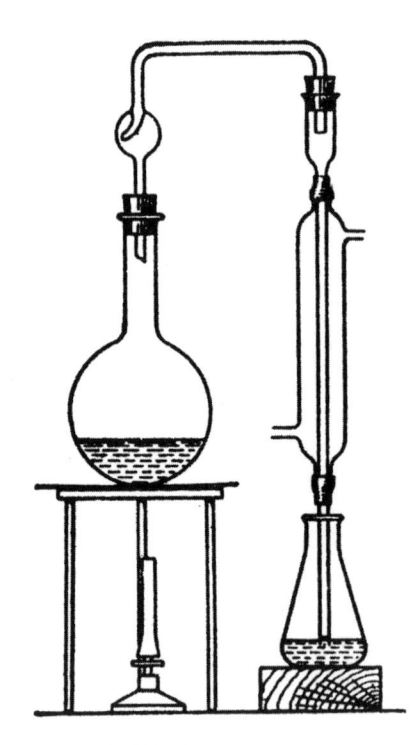

Distillation Apparatus

fumes of ammonia, perhaps laced with the nose-wrinkling stink of rotten eggs (hydrogen sulphide), may have brought back memories of school-day experiments. Whereas a pleasant smelling day in the laboratory could have recalled the agreeable aroma of ether, the distinctive pear-drop smell of amyl acetate, or the sweetish apple-like odour of acetone. On top of all this there were probably sounds, such as the gentle hiss of one or more of the gas supplied Bunsen burners, and also perhaps the regular drip of a tap or the frenzied bubbling of a heated retort or flask. Anyway, whatever the laboratory was like, here Stamford and Watson find Holmes to be the sole occupant, working at one of the benches. On hearing them approach, Holmes runs towards them clutching a test-tube in one hand. He is excited and calls out to them: *"I have found a reagent* [6A] *which is precipitated* [6A] *by haemoglobin* [6B]*, and by nothing else"* (A Study in Scarlet). He then proceeds to show Watson how it works, by pricking himself with a bodkin [6B] and using the resulting drop of blood for the test, which performs beautifully. At the same time he remarks that *"the old guaicum test* [6B] *was clumsy and uncertain."*

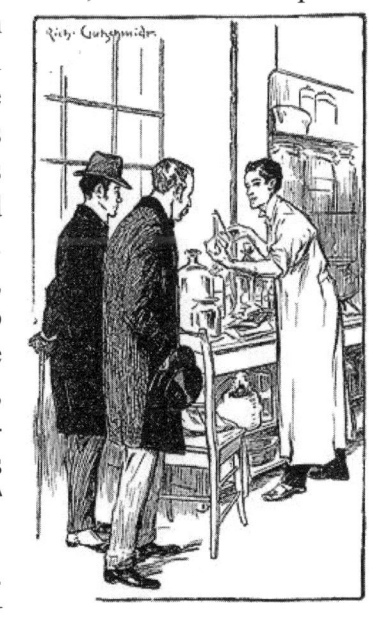

Holmes frequently turns to his test tubes and other chemical apparatus in several of the cases, and is careful before reaching an agreement to share rooms with Watson, to inform him that: *"I generally have chemicals about, and occasionally do experiments. Would that annoy you?" "By no means,"* replies Watson. Further discussion then ensues between Holmes and Watson about their mutual shortcomings and tastes, and with another meeting arranged between them, Watson and Stamford depart, with the former recording that: *We left him working among his chemicals.* So begins one of the most famous partnerships in fictional detective literature.

Both financial independence and chemistry are important to Holmes. In The Final Problem the reader learns that from some of his cases (including one concerning the royal family of Scandinavia, and another conducted on behalf of the French Republic), he has obtained enough money *"to concentrate my attention upon my chemical researches."*

Holmes' use of analytical chemistry [6B] certainly helps him to solve crimes, and at the same time increase his knowledge of this particular branch of science which so fascinates him. In this way he carries out valuable research, as exemplified by his experimental test for blood in A Study in Scarlet. By doing so, he hopes to push forward the boundaries of this new (in the 19th century) discipline, which today is known as 'forensic science' [6B].

Chemistry and biochemistry [6B] are essential aspects of forensic science, and various tests and methods have developed over the years. These include analytical, DNA [6B] and blood tests, as well as the microscopic and chromatographic [6B] examination of samples of substances and materials obtained from the scenes of crime. Although techniques have improved and many tests are now more sophisticated, forensic science still cannot solve every problem; though modern laboratory methods may well have obtained a result in The Speckled Band, when at the end of the investigation, Holmes comments on the cleverness of the murderer in *"using a form of poison which could not possibly be discovered by any chemical test"*.

As has been seen already, Holmes uses his chemistry for crime solving, but he also indulges in chemical experimentation, both as a hobby and as a form of relaxation: *Without his chemicals he was an uncomfortable man* (The Three Students). And in his own words: *"Well, I gave my mind a thorough rest by plunging into a chemical analysis"* (The Sign of Four). His keenness in experimenting sometimes means that he works all night, a procedure which Watson comments upon in The Copper Beeches: *Holmes was settling down to one of those all-night researches which he frequently indulged in, when I would leave him stooping over a retort and a test-tube at night, and find him in the same position when I came down to breakfast in the morning.*

However Holmes' experiments are not without their little mishaps, and can result in the occasional production of substances which no one in a normal domestic setting would wish to encounter. Gases, especially those with obnoxious smells, are a particular problem, and sometimes assail Doctor Watson and the long-suffering Mrs. Hudson, Holmes' housekeeper. In one story (The Sign of Four) Doctor Watson comments on this phenomenon: Holmes *busied himself all the evening in an abstruse chemical analysis which involved much heating of retorts and distilling* [6C] *of vapours, ending at last in a smell which fairly drove me out of the apartment. Up to the small hours of the morning I could hear the clinking of his test-tubes, which told me that he was still engaged in his malodorous experiment.* These *malodorous scientific experiments* are also mentioned in The Dying Detective.

METHODS AND EQUIPMENT

To carry out his experiments, Holmes had to acquire all the necessary apparatus comprising test tubes, beakers, bottles of chemicals, reagents for tests, flasks, pipettes [6C], Bunsen burners [6A], crucibles [6C], funnels, spatulas, tongs, balance [6C], and a retort for distillation.

His 'laboratory' is really part of his consulting room with a *chemical corner and acid-stained, deal-topped table* (The Empty House). It is mentioned several times in the stories, thus in The Musgrave Ritual Doctor Watson relates in his account that: *our chambers were always full of chemicals;* and at the beginning of The Dancing Men, when Holmes is working on a chemical analysis, he sits at the bench, bent over *a chemical vessel in which he was brewing a particularly malodorous product.* On this occasion when he turns round to look at Watson, he is holding in his hand a steaming test-tube, which he then places in a convenient rack nearby.

At the start of the account of The Naval Treaty, Holmes is carrying out another chemical investigation when Watson enters the room and stops to watch him at work: *A large curved retort was boiling furiously in the bluish flame of a Bunsen burner, and the distilled drops were condensing* [6C] *into a two-litre measure* [6C]. *He dipped into this bottle or that, drawing out a few drops of each with his glass pipette, and finally brought a test-tube containing a solution over to the table. In his right hand he held a slip of litmus-paper* [6D]. *"You come at a crisis, Watson," said he. "If this paper remains blue, all is well. If it turns red, it mean's a man's life." He dipped it into the test-tube and it flushed at once into a dull, dirty crimson.*

The "Gloria Scott" case informs us that Holmes' interest in chemistry is a long-standing one, and that in his student days he spent seven weeks of his vacation working on organic chemistry [6D] experiments, although no further details of these are given in the story. A more specific organic chemistry experiment though, is mentioned in The Sign of Four as Holmes tells Watson: *"When I had succeeded in dissolving the hydrocarbon* [6D] *which I was at work at, I came back to the problem."*

CHEMICALS AND CHEMICAL PRODUCTS

In several of the stories, certain chemical substances are mentioned. Thus in The Cardboard Box Holmes makes the point that the severed ears sent to Miss Cushing are surrounded by *rough salt* [6D] *of the quality used for preserving hides and other of the coarser commercial purposes.* He goes on to suggest that the ears probably did not come from a dissecting room where they would have been kept in preservative. *"Carbolic* [6D] *or rectified spirits* [6D] *would be the preservatives which would suggest themselves to the medical mind, certainly not rough salt."*

Another case (The Red-Headed League) refers to the fact that Vincent Spaulding has a white acid splash [6D] on his forehead, but we are not told which acid has caused this scar.

In The Blue Carbuncle a vitriol [6E] throwing incident is just one of a number of foul crimes carried out in order to obtain possession of the precious jewel of that name. Another vitriol attack occurs in The Illustrious Client, when Baron Gruner has vitriol thrown at his face by a revengeful Kitty Winter, a victim of his blackmail. Watson, who is present at the time, gives a graphic account of the terrible anguish caused by the burning acid: *The Baron uttered a horrible cry – a yell which will always ring in my memory. He clapped his two hands to his face and rushed round the room, beating his head horribly against the walls. Then he fell upon the carpet, rolling and writhing, while scream after scream resounded through the house.* Watson gives him first aid [6E], and later remembers that a footman came rushing into the room but suddenly fainted at the sight, *as I knelt by the injured man and turned that awful face to the light of the lamp. The vitriol was eating into it everywhere and dripping from the ears and the chin.*

Another common acid is mentioned in a brief reference in A Case of Identity when Watson calls at Baker Street and finds Holmes half asleep, with his used chemical apparatus all around, and with *the pungent cleanly smell of hydrochloric acid* [6E] *in the air.* The same story also has the following exchange between Watson and Holmes, with Watson the first to speak:
   "Well, have you solved it?" I asked as I entered.
   "Yes. It was the bisulphate of baryta [6E]*."*
   "No, no, the mystery!" I cried.
   "Oh, that! I thought of the salt that I have been working upon."
This brief misunderstanding by Holmes emphasizes the importance of his chemical work and how engrossed in it he could become to the

exclusion, albeit temporarily, of the case he was working on at the time.
A chemical paste daubed on a giant dog to achieve a 'flickering flame' effect has significant impact in one of the novels, and **The Hound of the Baskervilles** is probably the most famous story ever, to make use of luminosity. The terrifying appearance of the hound painted with this luminous paste [6F], strikes horror into the hearts of all who see it. With an excellent storyline, good characters, and a desolate Dartmoor setting, this superlative tale should give every reader much ghoulish delight. Especially notable are the fine graphic descriptions of the massive animal upon which the adventure is based. *A hound it was, an enormous coal-black hound, but not such a hound as mortal eyes have ever seen. Fire burst from its open mouth, its eyes glowed with a smouldering glare, its muzzle and hackles* [6F] *and dewlap* [6F] *were outlined in flickering flame. Never in the delirious dream of a disordered brain could anything more savage, more appalling, more hellish, be conceived than that dark form and savage face which broke upon us out of the wall of fog.*

Later when Holmes and Watson look at the body of the hound, Watson identifies the glowing substance as phosphorus. *"A cunning preparation of it"* adds Holmes. *"There is no smell which might have interfered with his power of scent."*

In the same novel Holmes detects the scent of white jasmine [6F] on the note sent to Sir Henry Baskerville by Mrs. Stapleton. It is here that Holmes tells us that he is an expert on scents, and claims that: *"There are seventy-five perfumes, which it is very necessary that the criminal expert should be able to distinguish from each other."*

In one adventure (**The Six Napoleons**), plaster of Paris [6F] is referred to as the material used to make busts of Napoleon, while in another (**The Sign of Four**), Holmes reveals that his monograph on tracing footprints contains notes on the use of plaster for preserving them.

For **The Copper Beeches** investigation Holmes has to tear himself away from experimenting with the acetones [6F] in order to travel to Winchester; and in **The Empty House** we learn more about his past laboratory work, as he recounts to Watson how he spent his time after surviving the momentous Reichenbach Falls struggle with Moriarty.

"I spent some months in a research into the coal-tar derivatives [6G]*, which I conducted in a laboratory at Montpellier* [6G]*, in the south of France."*

There are other references to coal-tar derivatives, which include creosote [6G], and to the fact that a carboy [6G] of this particular dark viscous liquid which is present in Bartholomew Sholto's chamber (equipped as a laboratory) has leaked (The Sign of Four), *for a stream of dark-coloured liquid had trickled out from it, and the air was heavy with a peculiarly pungent, tar-like odour.* As Bartholomew's murderer has stepped in the chemical, the creosote provides a useful trail across London, which Holmes and Watson follow with a dog named Toby, who has a keen nose for such outings. When Holmes dips a handkerchief in the creosote and pushes it under the dog's nose, the animal quickly takes up the scent. On one occasion the dog follows a false trail which takes them (with some hilarity on their part when the mistake is realized), to a timber yard, where a leaking barrel of creosote has recently been delivered. On retracing their steps, Toby regains the correct route and leads them on to a riverside wharf and the end of the trail for the time being.

Chloroform [6G] which has anaesthetic properties, is used in a kidnapping by criminals in The Disappearance of Lady Frances Carfax, but Holmes and Watson track down Lady Carfax and find the poor woman enclosed in a significantly macabre container – a coffin. *With a united effort we tore off the coffin-lid. As we did so there came from the inside a stupefying and overpowering smell of chloroform* [6G]. To restore the victim to consciousness, Watson administers ether [6G] by injection, and at last, with artificial respiration there is some flutter of life.

In another story (A Scandal in Bohemia) some simple observational detective work and his scientific knowledge, enable Holmes to rightly deduce that Watson is actively practising medicine again. He notices that the doctor smells of iodoform [6G] and has a black mark of silver nitrate [6H] on his right forefinger.

It should be noted that Holmes is not the only person in the stories to have an interest in chemistry. Bartholomew Sholto's chamber is also fitted out as a chemical laboratory and there is a good description of it: *A double line of glass-stoppered* [6H] *bottles was drawn up upon the wall opposite the door, and the table was littered with Bunsen burners, test-tubes, and retorts. In the corners stood carboys* [6G] *of acid in wicker* [6H] *baskets* (The Sign of Four).

NOTES 6A

Saint Bartholomew's Hospital (popularly known as Bart's) in the City of London is one of the oldest hospitals in Britain, and was founded by Rahere, an Augustinian monk, in 1123.

Retorts: Glass vessels used in the laboratory for the distillation of liquids.

The Bunsen lamp or Bunsen burner has been widely used over the years in laboratories to heat vessels holding solutions and solids. In Holmes' day it could readily be connected to, and operated from the household gas supply. The burner was developed in the laboratory of Robert Wilhelm Bunsen, Professor of Experimental Chemistry at Heidelberg University and was named after him; though it was probably designed by his technician, Peter Desdega and based on an earlier piece of apparatus invented by Michael Faraday. The Bunsen burner became popular as a laboratory heat source in the 1860's because its adjustable air valve allowed air to be supplied in a way that enabled better combustion of the gas, thereby creating a hotter flame.

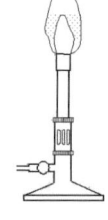

A reagent is a chemical material which will react with another chemical substance to which it is added, sometimes under the influence of applied heat, resulting in one or more chemical and/or physical changes.

Precipitated: The resultant effect of the separation of a solid from solution due to a chemical reaction or to the lowering of the temperature.

NOTES 6B

Haemoglobin is the red pigment found in the blood of vertebrates and comprises four polypeptide chains of approximately equal size and to each of which is attached a haem group. Haemoglobin functions by loosely binding with oxygen, which it carries round the body and also aids removal of waste carbon dioxide from the tissues.

Bodkin: A sturdy metal needle with a blunt tip and large eye to take a thick thread or tape.

The guaicum test uses guiacum resin (from *Guiacum officinale* or *Guiacum sanctum*) in an alcoholic solution with hydrogen peroxide. If these are mixed with a substance containing haemoglobin, a blue colour is obtained, thus indicating the presence of blood.

Analytical chemistry is the science of examining substances in order to determine and measure their component parts.

Forensic science: The practical application of scientific methods to provide data and hence evidence for use in criminal cases in court.

Biochemistry: The chemical aspects of biology controlling the lives of plants and animals, including human beings.

DNA, Deoxyribonucleic acid: The double helix structure of DNA was established by Francis Crick and James Watson in 1953. It is now thought that the DNA of each person is essentially unique and although a DNA test may be able to show the presence (or otherwise) of an individual at the scene of a crime, such evidence should not be used alone in order to convict an individual. DNA patterns can be complex, and human error may lead to poor test results which are not satisfactory for definite proof of a person's identity.

Chromatography is based on the work of Mikhail Tswett, who studied the separation of substances from solutions passing through columns of solid adsorbents. Chromatography and its variations, such as gas chromatography, are widely utilized in separating and measuring the constituents of forensic samples.

NOTES 6C

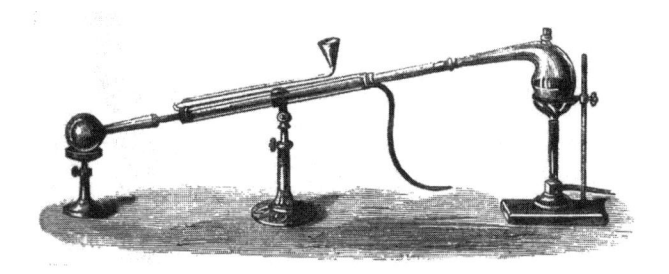

Distillation is a process in which a liquid is heated to its boiling point in a flask or retort and the vapour formed is condensed and collected in a cooled vessel. In the 19th century Liebig's water-cooled condenser became widely used, and by applying a vacuum to the distillation system it was possible to separate the volatile liquid components of a mixture by heating at much lower temperatures.

A pipette was basically a calibrated glass tube pointed at one end, which could be filled with a measured amount of a liquid. It usually had a rubber bulb at the other end which was used to draw up and discharge the liquid contents of the pipette. There was also a larger version which had a bulge in the centre of the tube. Suction of the liquid into the bulb was made by the mouth, and discharge of the liquid effected by gravity. Spring operated pipettes are now available.

Crucibles: In the laboratory these were usually of porcelain and were small round rimmed dishes used for ashing or fusing dry substances.

Balance: An accurate and delicate pivoted weighing machine, requiring individual weights to precisely measure dry chemicals, liquids in beakers and flasks, and residues after experiments.

Condensing: In this context, a vapour or gas which forms droplets on a cool surface, the droplets then coalescing to form a liquid.

2 litre measure: A glass container graduated to hold 2000 mls of a liquid.

NOTES 6D

Litmus paper: Usually strips of absorbent paper stained with the colouring agent, azolitmin, which comes from a lichen such as Lecanora or Rocella. The azolitmin is produced by fermentation of the lichen in ammonia, with potassium or sodium carbonate added. The action of an acid on the blue azolitmin turns it red, and an alkali added to this will turn it back to blue. The litmus paper, produced in both blue and red colours, could be used as an 'indicator' for acids and alkalis. There were also papers which could indicate degrees of acidity and alkalinity (pH values). In 1664, Robert Boyle had suggested that plant extracts could be used as chemical indicators in his *Experimental History of Colours*. However knowledge of the alkali/acid effect of plants was known for many years before this.

Organic chemistry: The study of organic compounds which are substances containing carbon, usually in combination with hydrogen and oxygen and certain other elements.

Hydrocarbons are compounds containing only carbon and hydrogen. They are often sub-divided into aliphatic (acyclic) and cyclic compounds. Examples of aliphatic hydrocarbons are butane and ethylene; and examples of cyclic hydrocarbons are cyclohexane and benzene.

Rough salt: Sodium chloride is the chemical name for salt. Rough salt (probably rock salt or marine salt in this context) would act as a crude preservative for organic matter. For example, in the past sailors found that salted beef remained edible for a considerable time during long voyages.

Carbolic: Phenol or carbolic acid was often used as a disinfectant.

Rectified spirits: An azeotropic (constant boiling) mixture containing distilled ethyl alcohol and 4% - 5% water.

Acid splash: It could have been caused by any strong corrosive acid as care has to be taken with all acids, and particularly with sulphuric acid. When diluting the latter with water, it is always necessary to remember that the acid should be poured into water and not the other way about, as this can be dangerous due to the liberation of much heat which causes the acid to 'spit' droplets of corrosive fluid which can burn and scar the skin.

NOTES 6E

Vitriol (sometimes called **oil of vitriol**) is another name for sulphuric acid, which is a clear, oily and highly corrosive chemical in its concentrated form. Vitriol used in crime: In Victorian and Edwardian London, the act of mutilating victims with acid appears to have been a more frequent occurrence than it is now. Today it is comparatively rare to read of such a case in the press, perhaps partly because methods of crime change like everything else. In the 19th and early 20th centuries it was easier to obtain dangerous chemicals. Most pharmacies (chemists then) would supply small quantities of certain acids over the counter, and these could also often be obtained from local hardware stores, and later from radio shops and garages (for batteries or accumulators which required sulphuric acid).
First Aid: (For details of Watson's treatment of the burning effect of the acid see notes to Chapter 2).

Hydrochloric acid: A common acid, sometimes called Spirit of Salt, this name being derived from its method of preparation by the action of sulphuric acid on salt (sodium chloride). The hydrogen chloride gas generated is dissolved in water to produce hydrochloric acid, a clear colourless liquid which may burn the skin on contact.

Bisulphate of baryta. Baryta is the common name for barium oxide, BaO. When heated with concentrated sulphuric acid it forms barium sulphate and barium bisulphate, $Ba(HSO_4)_2$.

NOTES 6F

Luminous Paste: When Conan Doyle wrote The Hound of the Baskervilles, a luminous mixture at that time may possibly have contained phosphorus, the luminescence resulting from an oxidative reaction of the phosphorus. Alternatively, a chemical such as impure barium sulphide or calcium sulphide may have been used, with the phosphorescence due to the presence of traces of heavy metals in the compound.

Hackles: The hairs on a dog's neck or back which may stand up when the dog is upset.

Dewlap: A fold of loose skin descending from the throat.

Jasmine: The perfume comes from the jasmine plant, of which there are several varieties. Both *Jasminum officinale* which has white flowers, and *Jasminum nudiflorum* which has yellow flowers, are highly scented.

Plaster of Paris is produced by heating gypsum at 120-130°C and is a partially hydrated form of calcium sulphate. Plaster of Paris sets very rapidly when water is added, and gives off heat at the same time. It is a very convenient product which can be used for making casts, sculptures, mouldings and for setting broken limbs, as well as a filler to repair cracks and holes in walls. The name originates from the prolific gypsum quarries at Montmartre (Paris).

Acetone (dimethyl ketone) is a commonly produced and used industrial solvent. It is a colourless liquid, volatile and inflammable, and with a pleasant slightly fruity smell. It readily mixes easily with water or ethyl alcohol. It is used as the chemical precursor of a number of useful compounds such as isopropanol. The human body produces small amounts of acetone which increase with certain diseases such as *diabetes mellitus*.

NOTES 6G

Coal tar derivatives: Today natural gas has widely replaced coal gas, and therefore coal tar, a by-product of coal gas generation, is not generally available. However in Victorian times large amounts of coal tar were produced and were utilized as the source of numerous useful substances such as benzene, naphthalene, creosote oils, carbolic acid (phenol), naptha and the cresols. Medicinal products as well as dyestuffs and other valuable products were also derived from coal tar constituents.

University of Montpellier: We can only conjecture where Holmes must have carried out his researches. The most likely location is at the University of Montpellier, where the faculty of medicine has had a long and notable history.

Creosote: an oily, viscous tarry substance recovered from coal tar, has been used for treating timbers to protect them from wood rot and wood pests. Creosote was studied in the 1830s by the German industrialist, Karl Baron von Reichenbach. The name Reichenbach will be familiar to all enthusiasts of the Sherlock Holmes stories, as the waterfall site of the life-and-death struggle between the famous detective and his opponent, Professor Moriarty.

Carboys are giant bottles of liquid chemicals, usually surrounded by a protective metal cage or one of wickerwork. Sometimes the bottles were packed in straw to prevent breakages.

Chloroform (trichloromethane) is a colourless, volatile liquid with a sweet odour. It was discovered in 1831 by Justus von Liebig (although Samuel Guthrie and other scientists also discovered it about the same time). It was first used as an anaesthetic in 1847 by the Scottish obstetrician, Dr. James Young Simpson, but its early use was opposed by both medical and religious factions. However it soon became acceptable after it had been administered to Queen Victoria.

Ether (diethyl ether) is a highly volatile and flammable colourless liquid with a very pleasant smell. It has a long history but was not used as an anaesthetic until 1842. When injected, ether acts as a powerful stimulant to the heart.

Iodoform (tri-iodomethane) is a yellow, crystalline solid and was used as an antiseptic.

NOTES 6H

Silver nitrate (or lunar caustic) often provided in stick form, is an important silver salt which was used in Doctor Watson's day as an antiseptic for burns, for treating warts and for cauterization. Silver nitrate is prepared by dissolving metallic silver in hot dilute nitric acid, evaporating the solution and allowing the colourless crystals of silver nitrate to separate out. A solution of silver nitrate has an interesting property: it can be used as a black indelible ink, because although normally colourless, it decomposes on contact with organic matter; hence Watson's black stained finger is easily visible. Silver nitrate is also used in photography (see Notes 9B).

Glass stoppers were used in the bottles of chemicals and acids in the laboratory because generally they were resistant to chemical attack, unlike the more common cork and rubber stoppers which could disintegrate or perish within a short time.

Wicker baskets are made of plaited lengths of willow shoots or branches, and among other applications were used as a protective covering for carboys. See note on carboys under 6G.

Pharmaceutical Chemistry: See Chapter 2

Photographic Chemicals: See Chapter 9

The Three Students

7 THE CRYPTIC MAZE

CALLIGRAPHY AND CRYPTOGRAPHY

For the detectives of the time, the scrutiny and comparison of handwriting and the identification of writing papers by appearance, quality, and watermark provided essential knowledge required for the detection of forgeries. Today, with the widespread use of printed text from computers, word processors and typewriters, handwriting is not so common on documents, but comparisons may still be required where signatures are involved or the papers are historic and hand-written. However, present day experts now believe that a message or document will contain an individual writer's style almost unique to that person. For example this can be in spelling, the use of certain words, or in the method of construction of the sentences. Such tests can be applied to printed documents considered to originate from a specific person, and so provide proof of authenticity or not, as the case may be.

Analysis of types of paper, quality, watermarks and residues from fingers and thumbs are all grist to the detection mill of Sherlock Holmes in his search for the truth, as will be shown in the sections below.

FORGERY

At the conclusion of The Three Garridebs mystery, Holmes and Watson are led by the unsuspecting 'Killer' Evans to the counterfeiters' cellar.

Our eyes fell upon a mass of rusted machinery, great rolls of paper, a litter of bottles, and, neatly arranged upon a small table, a number of neat little bundles. Cornered, Evans confirms that the machine is Prescott's, a forger of that name who was killed by him some time ago. *"Those bundles on the table are two thousand of Prescott's notes worth a hundred each and fit to pass anywhere. No living man could tell a Prescott from a Bank of England* [7A]*, and if I hadn't put him out he would have flooded London with them."*

Holmes refers to a forged cheque in The Mazarin Stone when he reads from his notebook some of the criminal activities of Count Sylvus. He tells the Count that: *"You are all here – every action of your vile and dangerous life. Here is the forged cheque on the Crédit Lyonnais* [7A]*."*

A vicar in The Retired Colourman claims that a telegram purportedly sent by him, *"is a scandalous forgery, the origin of which shall certainly be investigated by the police"*. And who is the forger? None other than Sherlock Holmes himself, who arranged for an agent to send the offending telegram from the village where the outraged vicar lived. If it had happened today, a Sunday tabloid would probably have run the headline: 'VICAR DUPED BY FAMOUS DETECTIVE'.

HANDWRITING

Among Holmes' accomplishments is a comprehensive knowledge of various forms of handwriting, a subject on which he has even published a monograph, as we learn in The Hound of the Baskervilles. In this adventure he quickly identifies the manuscript which Doctor Mortimer is carrying as one dating from about 1730. Holmes takes the historic document from him and not very professionally, flattens it on his knee. *"You will observe, Watson, the alternative use of the long ' s' and the short. It is one of several indications which enabled me to fix the date."* This manuscript relates the account of the mythology of the hound which has menaced generations of the Baskerville family.

Only a small number of the detective's cases are involved with people's handwriting. In **The Stockbroker's Clerk** a copy of Hall Pycroft's writing is obtained by the roguish Mr. Pinner on behalf of his dishonest brother. This brother needs it so that he can pose as Pycroft at a City stockbroking company.

In another case (**The Reigate Puzzle**) when a sample of handwriting is required by Holmes, he uses a slight deception to obtain it. He wants to compare the handwriting of old Mr. Cunningham with the writing on a note sent to William Kirwan, his coachman; so he gets the man to make a written alteration to a newspaper notice that he has prepared. By perusal of the changes Holmes deduces that the original note has been written by two people, the elderly Mr. Cunningham and his son, each taking his turn to inscribe the words. *"My dear sir,"* cried Holmes, *"there cannot be the least doubt in the world that it has been written by two persons doing alternate words."* In due course, Holmes points out the differences between the weaker and the stronger handwriting in the note, and suggests that the man with the stronger hand was the ringleader, writing all his words first and *"leaving blanks for the other to fill up. These blanks were not always sufficient. I have no doubt at all that a family mannerism can be traced in these two specimens of writing."* Holmes also expresses the opinion that handwriting experts would have found *"twenty-three other deductions"*. An exaggeration perhaps?

Handwriting is also important in another case (**The Valley of Fear**). Holmes receives a letter, and after removing the message inside, carefully studies the envelope. *"It is Porlock's writing,"* said he thoughtfully, *"I can hardly doubt that it is Porlock's writing, though I have only seen it twice before."* When a second letter arrives from Porlock, he identifies it immediately: *"The same writing,"* remarked Holmes, as he opened the envelope.

A comparison of writing in **The Sign of Four** reveals that all the packages containing pearls sent to Miss Morstan, have had the addresses on them written by the same person. She has also received a mysterious letter. First of all Holmes asks her for *"The envelope, too, please. Post-mark, London, S.W. Date, July 7. Hum! Man's thumb-mark on corner – probably postman. Best quality paper* [7A] *. Envelopes at sixpence* [7A] *a* *packet. Particular man in his stationery. No address."* Holmes also praises Miss Morstan as a model client because she has kept all the wrappings or labels (we are not told which) from the packages sent to her.

Mrs. Warren, a landlady brings Holmes an unusual case concerning a shy retiring lodger and some mysterious notes (The Red Circle). When one of the notes, with the letters printed in pencil, is scrutinized, Holmes is puzzled: *"Why print? Printing is a clumsy process. Why not write?"* Watson suggests that the reason is to conceal the lodger's handwriting. On the face of it a fairly obvious deduction, which Holmes could surely have made himself.

PAPER, PEN AND INK

In The Sign of Four adventure, Holmes identifies the paper used for a document which Miss Morstan hands to him: *Holmes unfolded the paper carefully and smoothed it out upon his knee. He then very methodically examined it all over with his double lens. "It is paper of native Indian manufacture," he remarked. "It has at some time been pinned to a board."* He then describes the contents of the document and concludes by saying: *"No, I confess that I do not see how this bears upon the matter. Yet it is evidently a document of importance. It has been kept carefully in a pocket-book, for the one side is as clean as the other."* However it turns out that it does bear upon the matter, and Holmes is correct about the importance of the document.

Mycroft, Sherlock Holmes' brother, has a part to play in the case of The Greek Interpreter. He places an advertisement in a newspaper requesting the whereabouts of a certain Greek man and woman. On receiving the reply, which conveniently gives a lead in the case, and by analysing the paper on which the answer has been sent, he makes a deduction. He observes that it was *"written with a J pen [7A] on royal [7A] cream paper by a middle-aged man with a weak constitution."* Mycroft undoubtedly has similar skills to those of his better known brother, though Holmes thinks that Mycroft's talents are in fact superior to his own.

A 'J' pen is also mentioned in The Cardboard Box: *"Done with a broad-pointed pen, probably a J, and with very inferior ink."* remarks Holmes.

In The Man with the Twisted Lip Holmes is amazed when Mrs. St. Clair tells him that she has received a letter from her husband, whom Holmes believes is dead. He is eager to look at this missive, and rather impolitely, snatches it from her in order to examine it closely. Watson, looking over his shoulder, sees that: *The envelope was a very coarse one. "Coarse writing,"* murmured Holmes. *"Surely this is not your husband's writing, madam."* Mrs. St.Clair agrees that it is not, but that the enclosure is. From the envelope and ink Holmes deduces that the writer had to inquire about the address, as the black ink used for the name has dried, but the actual address has been blotted [7A], revealing a greyish shade of ink.

The Valley of Fear mystery has Holmes examining a cryptic clue written on a piece of card. *"It is rough cardboard. Have you any of the sort in the house?"* he asks. *"I don't think so"* replies Ames, the butler. *Holmes walked across to the desk and dabbed a little ink from each bottle on the blotting paper. "It has not been printed in this room,"* he said; *"this is black ink and the other purplish. It has been done by a thick pen, and these are fine,"* he says, referring to the pens on the desk.

When Sir Henry Baskerville receives a letter on a half-sheet of foolscap paper [7A], consisting of cut-out and pasted printed words, Holmes is certain that the printed pieces have come from *The Times* newspaper [7B]. Doctor Mortimer is impressed by Holmes' recognition of the correct typeface of *The Times*, but Holmes dismisses it airily, confessing: *"that once when I was very young I confused the Leeds Mercury with the Western Morning News."* Continuing, he declares *"the detection of types is one of the most elementary branches of knowledge to the special expert in crime,"* and that *"a Times leader is entirely distinctive."* Holmes also notes that the printed words have been fixed to the paper using gum [7B] and not paste [7B]. There are other clues which Holmes picks up on. An interesting one is that the address on the envelope has been printed in a rough hand, but as Holmes remarks, *The Times* [7B] is invariably read only by the highly educated. Holmes suggests that the writer was an educated person trying to pose as one with lower educational ability. He also points out that *the pen had spluttered twice in a single word, and had run dry three times in a short address, showing that there was very little ink in the bottle.* From this, Holmes surmises that the writer has used the ink and a pen provided by a hotel. This whole sequence of deductions is most impressive. (The Hound of the Baskervilles)

A different newspaper is mentioned in another story (The Bruce-Partington Plans). Holmes recognizes the one from some discovered news cuttings which consist of *"a series of messages in the advertisements of a paper. The Daily Telegraph* [7B] *agony column* [7B] *by the print and paper."*

In A Case of Identity it is clear to Holmes that Miss Mary Sutherland *had written a note before leaving home, but after being fully dressed.* Although the torn forefinger of her right glove has been noticed by Watson, Holmes tells him that *"you did not apparently see both glove and finger were stained with violet ink. She had written in a hurry, and dipped her pen too deep."*

WATERMARKS

A watermark [7B] can be observed in some papers when they are held up to the light. The mark is usually a faint design indicating the manufacturer of the paper. It can also be used as a security aid, as in banknotes and other important documents. Although Holmes invariably looks for a watermark in the documents he scrutinizes, he does not always find one. Papers lacking watermarks are found in The Man with the Twisted Lip, and again in The Hound of the Baskervilles, where a half-page of foolscap paper has been used for a strange warning letter to Sir Henry Baskerville.

When Holmes receives a letter on pink-tinted notepaper (A Scandal in Bohemia), he asks Watson what he makes of it. Watson suggests that the writer *"was presumably well-to-do,"* and that *"such paper could not be bought under half a crown* [7B] *a packet. It is peculiarly strong and stiff"* he adds. Holmes of course has deduced a great deal more from it. He tells Watson to hold the paper up to the light, which he does, and sees *a large "E" with a small "g", a "P", and a large "G" with a small "t" woven into the texture of the paper.* Holmes explains the meaning of the letters and points out that the letters 'Eg' in the watermark indicate that it comes from Egria [7B] *"in a German-speaking country – in Bohemia, not far from Carlsbad* [7B]*".* It reveals that the paper is Bohemian.

TYPEWRITER

"It is a curious thing," remarked Holmes, *"that a typewriter has really quite as much individuality as a man's handwriting. Unless they are quite new, no two of them write exactly alike. Some letters get more worn than others, and some wear only on one side."* (A Case of Identity) Holmes then tells Mr. Windibank that *"I think of writing another little monograph some of these days on the typewriter and its relation to crime."* By comparing the four typed letters which had been received, with typed material from Mr. Windibank's office typewriter, and with the aid of other clues, Holmes solves the case.

CRYPTOGRAPHY

Cryptography [7B] is the science of creating ciphers and solving them. There are two stories (The Dancing Men and The Valley of Fear) in the canon which include cryptography as an important element of the cases, but The Sign of Four also has a brief reference when Holmes suggests that a cryptogram to solve would stimulate his mind, which otherwise would stagnate.

The Dancing Men is a fascinating case and involves a pictographic hieroglyphic cryptogram (try saying that quickly), which Holmes has received through the post. It has been sent to him by Hilton Cubitt, and when Watson first sees it, initially he thinks it is a child's drawing. When Hilton Cubitt arrives at the consulting rooms, Holmes gives his opinion of the 'drawing'. *"At first sight it would appear to be some childish prank. It consists of a number of absurd little figures dancing across the paper upon which they are drawn."* Later in the story, Holmes informs Inspector Martin that: *"I am fairly familiar with all forms of secret writings, and am myself the author of a trifling monograph upon the subject, in which I analyze one hundred and sixty separate ciphers."*

The Valley of Fear opens with Holmes receiving a message in cipher from one, Porlock, who we suppose has links with the infamous Professor Moriarty, or the 'famous scientific criminal' as Watson calls him. As everyone knows, to easily decipher a coded message, it is essential to have the key to the cipher. Holmes at this stage does not have it and believes it comes from a book: but all is not lost, as Holmes points out to Watson. The latter questions the logic of sending a message without indicating the book on which the message is based. Holmes answers this by suggesting that if Watson was to send such a message, his 'innate cunning' *"would surely prevent* [him] *from enclosing cipher and message in the same envelope"* in case it should go astray and fall into the wrong hands. True enough, by the second post comes the letter which Holmes expects to contain the title of the book with which to

Holmes Studies the Cipher

decipher the code. However, his hopes are dashed when the letter informs him that the writer, Porlock, will go no further with the matter as he is under suspicion from Moriarty. While Watson ruminates on the lack of information to solve the problem, Holmes applies his mind to actually solving the riddle. From reason he knows and tells Watson that: *"This man's reference is to a book. That is our point of departure."* Holmes' deductive powers are now at full strength as he speculates that the book in question must be a large one, as page 534 is mentioned at the start of the message. This is followed by 'C2', which Watson thinks may refer to 'chapter two', though Holmes demolishes this idea by saying *"if page 534 only finds us in the second chapter, the length of the first one must have been really intolerable."* Some plain speaking by Holmes but it holds some humour. At last the penny drops and Watson realizes it refers to 'column'. The reasoning in this vein continues. It cannot be the Bible, because there are too many different editions for the references to match. It is not Bradshaw [7C] because, as Holmes puts it beautifully. *"The vocabulary of Bradshaw is nervous and terse, but limited. The selection of words would hardly lend itself to the sending of general messages."* The dictionary is similarly dismissed. At last they hit on the solution of an almanac. Holmes tries Whitaker's Almanack [7C] but is defeated until he realizes that he is consulting the new edition, and not the previous one which Porlock must have used. Thus the cryptogram is solved. Watson remarks on the strangeness of the message, but Holmes points out the difficulties. *"When you search a single column for words with which to express your meaning, you can hardly expect to get everything you want."*

In the same story (The Valley of Fear) there is a second example of cryptography, which has the operator at the telegraph bureau [7C] in Hobson's Patch, telling McMurdo that Birdy Edwards, alias Steve Wilson, is sending coded messages every day on the pretence that he is a reporter for a New York paper, while all along he is a detective for the Pinkerton agency [7C].

NOTES 7A

Bank of England [note]: That is, an authentic note issued by the Bank of England.

Crédit Lyonnais: A large French bank, which has its headquarters in Lyons.

The best quality paper was made from rag, rather than woodpulp or plant material.

Sixpence = 6d. = 2½ pence in present currency.

'J' pen: Holmes points out that it is a 'broad-pointed pen'. The 'J' indicates the grading of the writing implement.

Royal paper: Standard size royal paper was 20 x 25 inches (approximately 51 x 63.5 cms).

Blotting Paper: Before blotting paper was introduced, a very fine powder (micaceous haematite) known as pounce was sprinkled on freshly written documents in order to dry the ink. The superfluous powder was then blown away, leaving the surface of the writing dry.

Foolscap paper: This unusual name is for a standard printing paper of 13½ x 17 inches (approximately 34.5 x 43 cms), and a drawing paper with slightly smaller dimensions. Commonly used typing paper with the same name measured 8 x 13 inches (approximately 20.5 x 33 cms). The name originated because a watermark of a fool's (or jester's) head with cap was widely used on sheets of the paper from early times.

NOTES 7B

The **gum** used for this purpose could have been gum arabic, which comes from Acacia trees.

Paste: A homemade variety could be made with flour and water.

***The Times* newspaper** started publication in 1785 as *The Daily Universal Register* by John Walter, and at one time was known as *The Thunderer* because of its forceful opinion. It became renowned for the quality and accuracy of its reporting and was widely read by the so-called upper classes, and also perhaps by those who aspired to a higher social standing.

The Daily Telegraph commenced publication in 1855.

Agony Column: A newspaper column for personal advertisements.

Watermark: A faint impressed design in paper usually indicating the manufacturer. Watermarks can now be investigated by means of radiography or digital imaging technology.

Half a crown = 2s.6d. = 12½ pence in present currency.

Egria is probably Eger (Cheb), a town which was then on the border of Bohemia.

Carlsbad or Karlsbad (now Karlovy Vary) was in Czecho-Slovakia. It is renowned for its hot springs of sodium sulphate (Glauber's Salts).

Cryptography: The construction and solving of codes and ciphers used for messages where security and secrecy is essential.

NOTES 7C

Bradshaw was a popular commercially published railway timetable guide covering the whole of Britain. It was started in Manchester in 1839 by publisher George Bradshaw and ceased publication in 1961. As the railway network was then vast, the timetable could be used to plot complicated journeys using both main and branch lines, with changes of train at as many large stations and small junctions as required.

Whitaker's Almanack is still published to-day. It is a comprehensive compendium of facts and figures about Britain and the world. It was founded in 1868 and first issued by London publisher, Joseph Whitaker in 1869.

Telegraph bureau: An office from which telegraphic messages (or telegrams) could be sent, for which a form had to be completed. Incoming messages were also received here, and after transcription sent by courier to the recipient's address. In the early days of the system, an operator used a keypad to transmit the messages in plain language or code (often Morse). At the receiving end needle pointers spelt out the letters for incoming communications, or a moving pen or inked wheel recorded them on a paper strip. On some systems a skilled operator could transcribe a message directly from a Morse sounder. There were several developments as demand for the service grew. Undersea cables were installed for overseas communication, and automatic systems were introduced using punched-hole tape. Pneumatic tubes linking local offices, teleprinters, and duplex systems gradually speeded up the service provided. Eventually the telephone took over.

The Pinkerton National Detective Agency was founded in Chicago in 1850 by Allan Pinkerton (1819-84), a Glaswegian by birth, and E. G. Rocker. The agency solved many sensational crimes and in time became world famous. The 'mugshot' or facial photograph of criminals was started by the Agency. It also used the Bertillon system of classification (see Chapter 8). Conan Doyle knew the founder's son, William Pinkerton.

The Naval Treaty

8 THE UNIVERSE AND NUMBERS

ASTRONOMY, METEOROLOGY and MATHEMATICS

ASTRONOMY

Holmes makes out that he has little interest in astronomy when Watson raises the question of the solar system in A Study in Scarlet. Holmes proclaims: *"you say that we go round the sun* [8A]. *If we went round the moon it would not make a pennyworth of difference to me or to my work"*. Then in The Hound of the Baskervilles, Watson remarks in a letter to Holmes: *"I can still remember your complete indifference as to whether the sun moved round the earth or the earth round the sun"*. However in the case of The Greek Interpreter, Holmes discusses the causes of the change in the obliquity of the ecliptic [8A].

Mr. Frankland of Lafter Hall [8A] is an amateur astronomer. He has an excellent telescope [8A] and uses it to sweep Dartmoor while trying to detect the escaped convict from the prison [8A] at Princetown (The Hound of the Baskervilles). (See also the chapter on Optics.)

In The Valley of Fear Inspector MacDonald describes how he once had a conversation with Professor Moriarty about eclipses [8A]. *"How the talk got that way I canna think; but he had a reflector lantern and a globe* [8A], *and made it all clear in a minute. He lent me a book, but I don't mind saying that it was a bit above my head."* Before this though, we are told by Holmes that Moriarty, his declared enemy, is *"the celebrated author of The Dynamics of an Asteroid* [8B] *– a book which ascends to such rarefied heights of pure mathematics* [8B] *that it is said that there was no man in the scientific press capable of criticizing it"*.

An unusual astronomical reference is made in The Stockbroker's Clerk, where Conan Doyle uses a simile to describe the look on Holmes' face as being *like a connoisseur who has just taken his first sip of a comet vintage* [8B]. Another instance is when Holmes makes the comment: *"A planet might as well leave its orbit* [8B]*,"* when he receives a telegram from Mycroft, his brother, informing him that he immediately intends to visit Baker Street. So extraordinary is such an event, that Holmes uses this expression (The Bruce-Partington Plans).

METEOROLOGY

It was a wild, tempestuous night. Outside the wind howled down Baker Street, while the rain beat fiercely against the windows. (The Golden Pince-Nez) This excerpt is just one of many that describe the prevalent weather conditions of fog, rain, wind, frost and snow encountered by Holmes and Watson throughout the canon.

Fog is the weather condition that readers frequently associate with the stories, but it is only found in a few of the mysteries. The metropolis at that time was often foggy because of pollution from the many smoking chimneys in the city, and Baker Street suffered as much as any other part of the area. The following passage appears in The Copper Beeches: *A thick fog rolled down between the lines of dun-coloured houses, and the opposing windows loomed like dark, shapeless blurs, through the heavy yellow wreaths.*

Fog also affected the countryside as well, and lonely Dartmoor, the setting for The Hound of the Baskervilles, sometimes suffered from banks of fog descending and blotting out familiar landmarks to confuse and misdirect anyone who did not know the area well. The following quote is an example. *Over the great Grimpen Mire there hung a dense, white fog.*

Snow, mainly because of its ability to 'record' tracks, footprints and bloodstains, proves useful in at least one of the cases. This is The Beryl Coronet, which opens with Holmes and Watson looking out from their window onto Baker Street. As Watson writes: *It was a bright, crisp February morning, and the snow of the day before still lay deep upon the ground, shimmering brightly in the wintry sun.* Later in the story this snow provides essential help to Holmes in solving the mystery.

Another wintry case, The Blue Carbuncle, again opens in Holmes' rooms at Baker Street, with a cold and recently arrived Watson telling us in his account. *I seated myself in his armchair and warmed my hands before his crackling fire, for a sharp frost had set in, and the windows were thick with ice crystals.*

In **The Boscombe Valley Mystery** Holmes makes an oblique reference to 'barometric pressure' [8B] in a reply to a comment from Inspector Lestrade. As Lestrade does not understand, Holmes adds: *"How is the glass* [8B]*? Twenty-nine* [8B]*, I see. No wind, and not a cloud in the sky".* Later in the same story Holmes is concerned that clues could be obliterated if the weather changed. *"The glass still keeps very high," he remarked, as he sat down. "It is of importance that it should not rain before we are able to go over the ground."*

Gales can affect animals and marine life as well as humans. In **The Lion's Mane** Holmes explains to Inspector Bardle that *"it may have been a south-west gale that brought it up."* Here the creature referred to is *Cyanea capillata* [8C], a jellyfish which has been found in a beachside pool.

At the beginning of the second part of **The Valley of Fear** a descriptive passage of the country around Vermissa [8C] announces that *it had been a severe winter and the snow lay deep.* At the end of the mystery Cecil Barker tells Holmes that a cable received by him from Mrs. Douglas, concerning her husband, briefly reports that: *Jack has been lost overboard in gale off St. Helena* [8C]. Like snow, rain could sometimes assist in the hunt for clues because muddy ground could preserve footprints and wheel marks. More about this can be found in Chapter 12.

An instrument to predict the weather will always be helpful, though not always accurate. At Mrs. Cecil Forester's house at Lower Camberwell [8C], Watson notices a barometer [8C] in the hallway (**The Sign of Four**).

MATHEMATICS AND MEASUREMENT

It has been said that Conan Doyle based his character of Moriarty on a real professor of mathematics named George Boole, but inverted the character of the man. Unlike Moriarty, Boole was not a criminal, but quite the opposite. A conscientious and brilliant mathematician, he was to become better known as the inventor of Boolean algebra, a fact which should be of interest to all computer users. Moriarty is also considered to be an outstanding mathematician, and in The Final Problem, Holmes refers to him as *"the retiring mathematical coach, who is, I dare say, working out problems upon a black-board ten miles away"*.

In The Sign of Four Holmes philosophizes on the nature of man, and quotes from a work by Winwood Reade [8C], saying that *"while the individual man is an insoluble puzzle, in the aggregate he becomes a mathematical certainty. You can, for example, never foretell what any one man will do, but you can say with precision what an average number [8C] will be up to. Individuals vary, but percentages [8C] remain constant. So says the statistician [8C]."* Towards the end of the adventure, when a boat chase takes place on the River Thames, Holmes has to decide which way to follow the steam launch Aurora. *"It is certainly ten to one [8C] that they go downstream, but we cannot be certain,"* he says.

Mathematics, or perhaps more accurately, geometry [8C], together with the position of the sun, play an important part in the solving of the mystery of The Musgrave Ritual. In the grounds of the house stands a magnificent oak tree. Holmes takes a sighting on this and measures the length of its shadow, using a length of fishing rod, a length of string knotted at each yard [8C], and a peg, as he explains to Watson in his narrative: *"The sun was just grazing the top of the oak. I fastened the rod on end, marked out the direction of the shadow, and measured it."* The length of the shadow of an elm tree is also required, but the tree has long since gone. Holmes asks Musgrave if the height of the missing elm is known, to which Musgrave replies: *"I can give you it at once. It was sixty-four feet."*

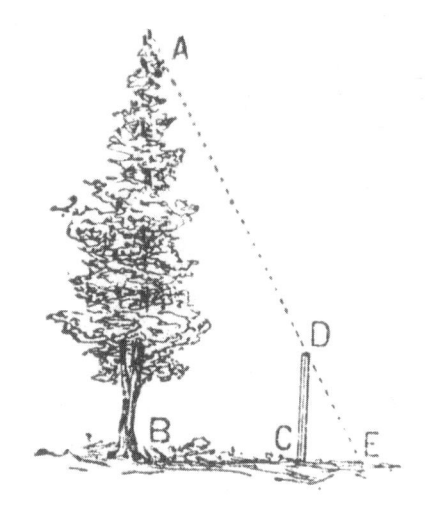

Holmes is surprised and asks *"How do you come to know it?"* Musgrave answers: *"When my old tutor used to give me an exercise in trigonometry* [8D]*, it always took the shape of measuring heights. When I was a lad I worked out every tree and building on the estate."* Having obtained the measurements, Holmes then makes the necessary calculations to find the correct spot at which to place his peg. *"From this starting-point I proceeded to step, having first taken the cardinal points* [8D] *by my pocket-compass,"* as he tells Watson, and so was able to follow the trail.

Further measurement, this time with a tape measure, is a necessary chore for Holmes at the scene of the murder of Drebber in A Study in Scarlet.

"We have boxed [8D] *the compass among us"* cries Inspector Bradstreet in The Engineer's Thumb, when the assembled party of investigators select every possible direction in which to search for the counterfeiter gang.

In Silver Blaze, Holmes calculates the speed of the train he and Watson are travelling in on their way to Devon. *"We are going well,"* says Holmes as he looks out of the window and glances at his watch. *"Our rate at present is fifty-three and a half miles an hour."*
"I have not observed the quarter-mile posts" Watson says.
"Nor have I. But the telegraph posts upon this line are sixty yards apart, and the calculation [8D] *is a simple one."*

On another train journey, this time to Woking [8D], Holmes discusses the Bertillon system of measurements [8D] with Watson. (The Naval Treaty)

When the binomial theorem [8D] is brought up in The Final Problem, Holmes tells Watson something about Professor Moriarty: *"At the age of twenty-one he wrote a treatise upon the binomial theorem, which has had a European vogue. On the strength of it, he won the Mathematical Chair at one of our smaller universities."*

NOTES 8A

Nikolaus Copernicus was responsible for the first detailed study of the idea that the earth revolved **round the sun**, and he worked on his theory for many years. However his views were at odds with those of the religious authorities who thought the earth to be at the centre of the Universe, and it was not until 1543, the year of his death, that his book on the subject, entitled *De Revolutionibus Orbium Coelestium*, appeared.

Obliquity of the ecliptic: The angle between the ecliptic and the celestial equator.

Lafter Hall: Lafter Hall is obviously a fictitious property. Everyone who reads The Hound of The Baskervilles will find that Mr. Frankland is an eccentric man who enjoys legal tussles with individuals or officials as a hobby. In frivolous mood could Conan Doyle have decided on the name of the property as a skit on 'Laughter Hall'!

Telescope: An instrument that produces enlarged images of distant objects. There are different types using combinations of lenses, or lenses and mirrors. Further information is available from scientific works and the many books on astronomy for which telescopes are essential.

Dartmoor prison: Situated at Princetown and built in 1806 for French prisoners of war. From 1850 it was converted to house convicts.

Eclipses: These usually refer to the Moon or Sun. One celestial body may pass in front of another and obscure it wholly or partially, so cutting off light by doing so.

Reflector lantern and globe: Moriarty obviously used the globe, a spherical representation of the world, and the lantern to demonstrate the effects of an eclipse.

NOTES 8B

Asteroids are small bodies or minor planets orbiting in space. In the solar system many are located in an asteroid belt between the orbits of Mars and Jupiter.

Pure mathematics is the study of mathematics and its branches on their own or in relation to one another. For example it includes arithmetic, algebra, geometry and calculus.

Comet vintage: A comet vintage is the wine produced in the year in which a notable comet appears or reappears. It is considered to be of a particularly fine quality. The best known of the comets is Halley's Comet which reappears every 76 years.

Orbit: A path through space of a star, comet, planet, or other heavenly body.

Barometric pressure: The measurement of the pressure of the atmosphere by means of a barometer.

The 'glass' is a colloquial name for the barometer.

Twenty-nine: The height in inches of mercury in the column of a barometer indicating the atmospheric pressure.

NOTES 8C

Cyanea capillata. **or the Lion's Mane** is a stinging jellyfish. In the past it was rare but could sometimes be found off the British coast. This jellyfish derives its name from its tentacles, which resemble a lion's mane and have been known to reach over 40 feet long on an exceptional specimen.

Vermissa: Untraced, probably a fictitious location in the U.S.A.

St. Helena: A British island in the South Atlantic Ocean.

Lower Camberwell is in South London.

Barometer: An instrument that measures atmospheric pressure and can forecast likely weather conditions.

William Winwood Reade (1838-75), was an English novelist, traveller and explorer in Africa. His book *The Martyrdom of Man* was published in 1872.

Average number: A mathematically calculated estimate obtained from a series of figures.

Percentage: A figure or quantity expressed as a proportion of 100 (i.e. 100%.).

Statistician: One who collects and analyses numerical data.

Ten to one: With the odds expressed as a ratio of 10 to 1 the likelihood of the Aurora going downstream is heavily weighted in favour of that direction.

Geometry. That branch of mathematics which deals with the dimensions and properties of lines, shapes and solids. The science goes back many years. Euclid (Third and Fourth centuries BC) and Pythagoras (Sixth century BC) are well-known early exponents.

One yard = 3 feet = circa 91.5 centimetres

NOTES 8D

Trigonometry: That branch of mathematics which deals with the solving of numerical problems relating to angles and triangles.

The cardinal points of the compass are North, South, East and West.

Boxing the compass is to explore (or state) all the points of the compass in turn and so return to one's starting point.

Calculation of the train speed. It is known that the telegraph poles are 60 yards apart and Holmes times the train between two adjacent poles, assuming that it travels at a constant speed. He must have found that it took 2.294 seconds in order to calculate the speed as 53.5 miles per hour. As his watch could not have given such an accurate figure, it is likely he timed the train past a series of poles before calculating the speed.

Woking is a town in Surrey, south of London.

The Bertillon system of measurement [named after the French criminologist Alphonse Bertillon (1853-1914)] who invented anthropometry, a method of measuring the parts of the human body. The system was of great use to police forces in the 19th century before the introduction of fingerprinting, which was simpler and more precise, and largely replaced the Bertillon system.

The binomial theorem. A binomial is an algebraic expression containing two terms, and the theorem provides a formula for calculating the expansion of a specific binomial raised to a given power as a series. The theorem is usually attributed to Isaac Newton, though the Chinese may have been the first to devise it some centuries earlier.

9 SEARCH FOR THE TRUTH

OPTICS AND PHOTOGRAPHY

Vision, which depends on light, is essential to everyone, but the naked eye often needs help in observing objects in greater detail. Optical instruments that utilize lenses and/or mirrors can aid the user to look at these objects and study them more clearly. Similarly poor sight can be aided by means of lenses in frames (glasses), pince-nez or even monocles.

MAGNIFIERS

A convex lens is necessary for magnification, and the use of a magnifying glass with such a lens plays an important part in many of the investigations of Sherlock Holmes. At least eighteen of the stories describe how Holmes produces his lens [9A] and diligently searches around the scene of a crime for clues, such as blood stains, bullet marks, and small traces of substances. Two examples are given here: *Suddenly he sprang up again and ran across to the opposite parapet, whipped his lens from his pocket, and began to examine the stonework* (The Problem of Thor Bridge). *Holmes had been examining the cover of the notebook with his magnifying lens. "Surely there is some discolouration here," said he. "Yes, sir, it is a blood-stain"* answers police inspector Stanley Hopkins (Black Peter).

A double lens, which can provide greater magnification, is used by him in The Sign of Four, as he scrutinizes a document. *Holmes unfolded the paper carefully and smoothed it out. He then very methodically examined it all over with his double lens.*

SPECTACLES

Glasses and spectacles, and occasionally pince-nez [9A] are mentioned in a number of the stories, and one of these actually has the title of The Golden Pince-Nez, where the pince-nez are a major factor in the crime and investigation.

Characters who wear standard frames and lenses include Sidney Johnson (The Bruce Partington Plans), and Nathan Garrideb who wears *large round spectacles* (The Three Garridebs). Doctor Mortimer favours gold-rimmed glasses (The Hound of the Baskervilles), as does John Scott Eccles, for whom Conan Doyle draws a highly effective word picture with this description: *From his spats to his gold-rimmed spectacles he was a Conservative, a churchman, a good citizen, orthodox and conventional to the last degree* (Wisteria Lodge).

Variations on the type of glasses worn by characters in the stories can be found. For example we have horn-rimmed [9A] glasses worn by Professor Presbury in The Creeping Man as described by Watson: *We were aware of a pair of keen eyes from under shaggy brows which surveyed us through large horn glasses.* Tinted glasses [9A] are worn by the sly Hosmer Angel, though the reason for this only becomes clear at the end of the case (A Case of Identity). The taxidermist [9A] is another wearer of spectacles, and again we have another excellent word picture: *Mr. Sherman was a lanky, lean old man, with stooping shoulders, a stringy neck, and blue-tinted glasses* [9A] (The Sign of Four).

An eyeglass, or monocle is worn by Colonel Ross, *the well-known sportsman,* in Silver Blaze; but poor Heidegger, a German school master is found dead wearing spectacles, one glass of which had been knocked out (The Priory School).

A field-glass [9A], which is often used by horse owners and trainers, is mentioned in Silver Blaze, the adventure about a missing horse.

The telescope [9A] has a part to play in The Hound of the Baskervilles. Mr. Frankland, the elderly resident of Lafter Hall [9A], a property which neighbours Baskerville Hall, is an amateur astronomer, and uses his telescope, which is situated on the roof, to observe the heavens. While doing so, he sees a boy (who turns out to be Cartwright), taking supplies to Holmes, who is living rough (but with clean linen and food delivered regularly by this youngster!) out on the moor. Whilst Frankland is telling Watson all about this, he happens to see the boy just at that moment, and they both rush upstairs to look through the telescope, which Watson describes as *a formidable instrument mounted upon a tripod.* Later when Watson himself finds Holmes on the Moor, it turns out that Holmes has already realized that he has been observed: *"The old gentleman with the telescope, no doubt. I could not make it out when first I saw the light flashing upon the lens."* Much later, when Holmes is looking at the portraits in Baskerville Hall, he asks: *"Who is the gentleman with the telescope?"* Sir Henry Baskerville informs him that it is Rear-Admiral Baskerville (no first name is given), an ancestor of the family.

MICROSCOPES

The microscope [9B], which gives an enlarged vision of objects placed under it, is perhaps of more importance than the telescope in the stories. Mounted and stained specimens on glass slides, small objects and examples of animal life in watch glasses or petri dishes [9B] can be placed under the instrument for clear and easy observation.

Holmes' own use of the instrument for the studying of forensic evidence occurs in Shoscombe Old Place. The story opens with Holmes bending over a low-power microscope. He looks up and invites Watson to have a peep. *"It is glue, Watson,"* said he. *"Unquestionably it is glue. Have a look at these scattered objects in the field!"* [9B] *I stooped to the eyepiece and focussed for my vision.* Holmes then describes what could be seen through the lens, namely: threads from a tweed coat, grey dust, epithelial scales [9B] and brown blobs of glue. Previously, using a microscope, Holmes had found zinc and copper filings in the seam of a cuff. He mentions that the police have *"begun to realize the importance of the microscope"*. A comment on its use in forensic science and aid to tracking down criminals.

In A Study in Scarlet Holmes observes that the microscopic examination of blood corpuscles *"is valueless if the stains are a few hours old."* In another story (The Three Garridebs) *the tall brass tube of a powerful microscope* is noticed by Holmes and Watson at the home of Nathan Garrideb, who has his own museum.

PHOTOGRAPHY

At the time that the first Sherlock Holmes stories were published in the 1880's and 90's, photography was still in its infancy, but there are references to this comparatively new science in The Red-Headed League, and again in A Scandal in Bohemia. Both these stories were first published in 1892 when there had recently been a number of important advances in photographic methods [9B].

In The Red-Headed League, Jabez Wilson, who operates a pawn-broking business [9B], tells Holmes that his assistant, Vincent Spaulding, is a keen photographer: *"Snapping away with a camera when he ought to be improving his mind, and then diving down into the cellar like a rabbit into its hole to develop* [9B] *his pictures."* At the end of the story, we learn that the photography was just a cover, giving Spaulding an excuse to spend long hours in the cellar where he was really excavating a tunnel under and into a neighbouring bank.

Another character who says that his hobby is photography is Mr. Rucastle, who wants to explain away a shuttered room (used as a dark room) in a wing of his house. (The Copper Beeches)

A photograph itself is the cause of the problem investigated by Holmes in A Scandal in Bohemia. This is in connection with an 'indiscretion' by a noble client with a tongue-twister of a name: Wilhelm Gottsreich Sigismond von Ormstein, Grand Duke of Cassel-Falstein and hereditary King of Bohemia. Irene Adler, a so-called 'adventuress' and the King's former companion, has retained in her possession a photograph of them together. She is threatening to ruin his Highness by releasing this to the strict family of his now intended fiancée when the couple's forthcoming betrothal is announced. Although the photograph is not successfully retrieved by Holmes, all ends reasonably well, but purely at the whim of Irene Adler. She cleverly manages to upstage the detective and flees the country, but sends him a note that informs him: *As to the photograph, your client may rest in peace.* One small point should perhaps be mentioned here. Holmes discovers that the photograph he was commissioned to find was of cabinet size [9C], and as he informs Watson: *"Too large for easy concealment about a woman's dress."*

Also found in some other cases are photographs of: a circus performer (The Veiled Lodger); Lady Hilda Trelawney Hope (The Second Stain); Brenda Tregennis (The Devil's Foot); a regal lady (Charles Augustus Milverton); and Josiah Amberley's wife (The Retired Colourman). In Silver Blaze Holmes uses John Straker's photograph for identification purposes, as he does in The Six Napoleons, having extracted the photograph of Beppo (an Italian thief and murderer) from the body of Pietro Venucci. We are told it was *"a snapshot* [9C] *from a small camera."*

In The Lion's Mane Holmes produces an enlarged photograph [9C] of the weals on MacPherson's body. *"This is my method in such cases,"* he tells Inspector Bardle. Here is an excellent example of a photographic record being obtained at the scene of a crime or incident for evidential purposes. Holmes' methods again should be applauded.

NOTES 9A

Lens: This is a convex magnifying lens, probably with a handle.

Pince-nez: From the French, meaning 'nose pinch'. These glasses had no side arms or ear-pieces, but instead a spring clip which pinched and kept them on the nose of the wearer.

Horn-rimmed: In Holmes' time, the frames would really have been made of horn from an animal, whereas today they would probably be manu-factured from plastic with the appearance of horn.

Tinted glasses: Spectacles with darkened or coated glass lenses to give the eyes protection from the sun and strong light. Tinted glasses can also be used as part of a disguise.

Taxidermist: A technician who prepares and preserves dead creatures for display.

Field-glass: This is either a small terrestrial telescope, or binoculars with an adjustable focus (usually called field glasses) popular with the racing fraternity, as well as seafarers, birdwatchers and anyone requiring a clear view of distant objects.

Telescope: An instrument that produces enlarged images of distant objects. There are various types using differing combinations of lenses, or lenses and mirrors, details of which can be found in astronomical and other science textbooks.

Lafter Hall is obviously a fictitious property. Everyone who reads The Hound of the Baskervilles, will find that Mr. Frankland is an eccentric man who enjoys legal tussles with individuals or officials as a hobby. In frivolous mood, could Conan Doyle have decided to name the property as a skit on 'Laughter Hall'?

NOTES 9B

Petri dish: A circular shallow dish of glass (or plastic now) with a loose-fitting lid.

Microscope: An optical instrument that magnifies objects and specimens placed under the objective lens and viewed through an eyepiece.

The 'field' here refers to the area or field of vision observable through the eyepiece of the microscope.

Epithelial scales: Dead tissue cells from a human or animal. In this case they may have been flakes of dandruff.

Pawnbroking business: A loan of money against pledged articles of value (such as jewellery, antiques etc.). An article is redeemed by the borrower on repayment of the loan with interest. Those items not redeemed are sold.

Developing and printing of photographs. The keen amateur would have developed and printed his own photographs, using a dark room, preferably equipped with a sink, although tanks could suffice. Other necessary equipment would have been developing dishes, printing frames, scales and chemicals, and an enlarger if the budget would stretch that far.

The discovery of celluloid proved to be a great advance for photographic techniques, and important developments for amateur photographers were the marketing of a suitable camera by Kodak from 1888 onwards, and the growing availability of paper-backed roll film.

During the period from 1887 to 1927 (when the last Holmes story appeared), some of the chemicals which may have been used for processing and were either available or introduced are as follows: Developers: – pyrogallol (pyrogallic acid); quinol (hydroquinone); rodinal (para-aminophenol). Silver nitrate was used in manufacturing photographic plates, films and papers and in the early wet plate process. In time silver halide light-sensitive papers were introduced. Silver bromide paper became available commercially from the 1880s and P.O.P. (Print-out-Paper) became popular from about 1891. The fixative known as hypo (sodium thiosulphate) was in common use.

NOTES 9C

A **cabinet photograph** is about 4.25 x by 6.5 inches (approximately 11 cms x 16.5 cms) in size, though this can vary slightly.

Snapshot: A quick and simple photograph taken using a hand-held camera.

Enlargement of a photograph: This requires a piece of apparatus called an enlarger, consisting of a stand with a calibrated column supporting a light source, bellows, condenser and enlarging lens. The image from the negative of the photograph to be enlarged is projected through the apparatus and on to light-sensitized paper.

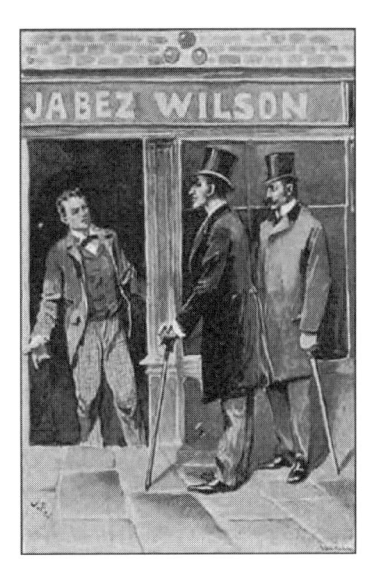

The Red-Headed League

10 TECHNOLOGY PLAYS ITS PART

ALL THINGS TECHNICAL

The stories abound in references to the machines, equipment and technological advances occurring at the time they were written, some of which have been covered in previous chapters; and the present review shows how extensively technology had started to enter into daily life in this period of 1890 to 1927.

TRANSPORT BY ROAD, RAIL AND SEA

The bicycle [10A] appears in several stories [10A], and is an important element in the solving of the problem in The Priory School where Holmes announces that: *"I am familiar with forty-two different impressions left by tyres"* when he and Watson come across some bicycle tyre tracks on the moor. Holmes identifies them as being created by Dunlop tyres [10A], whereas they are looking for the tracks of a bicycle with Palmer tyres. Holmes cleverly suggests that the cyclist was heading away from the school, because, as he tells Watson, *"The more deeply sunk impression is, of course, the hind wheel, upon which the weight rests. You perceive several places where it has passed across and obliterated the more shallow mark of the front one [10A]."* Later, Holmes finds the bicycle with the Dunlop tyres outside the Fighting Cock Inn. He can positively identify it from a patched tyre, the mark of which he saw in the tracks on the moor.

The Solitary Cyclist is more concerned with the rider than the machine, but the latter is inevitably an essential constituent of the yarn. In The

Five Orange Pips we learn that John Openshaw's father owned a small factory *"which he enlarged at the time of the invention of bicycling. He was the patentee of the Openshaw unbreakable tyre[10A]"*.

Horse-drawn vehicles, such as the Hansom cab [10A] play a part in many of the stories [10A]. One of these (The Final Problem) has Watson fleeing to the Continent with Holmes to escape the clutches of Professor Moriarty. Holmes leaves Baker Street before-hand and gives Watson detailed instructions on how to follow him. *"In the morning you will send for a hansom, desiring your man to take neither the first nor the second which may present itself. Into this hansom you will jump."*

Another story (The Red-Headed League) has Watson narrating his own actions. *I made my way across the Park, and so through Oxford Street to Baker Street. Two hansoms were standing at the door.* The reason is that Holmes has two visitors, Mr. Jones of Scotland Yard and Mr. Merryweather, Chairman of a City bank. After some discussion Holmes says: *"If you two will take the first hansom, Watson and I will follow in the second."*

The Growler [10A], also known as a four-wheeler or Clarence cab could be hired as an alternative to the Hansom. A Study in Scarlet has a beautiful scene where Mrs. Sawyer, an 'old crone', hires a passing four-wheeler. To follow her, Holmes hitches an unorthodox lift on the back of the cab. Later he ruefully describes to Watson what happened when it arrived at the destination given to the driver by the old lady. *"I saw the cab pull up. The driver jumped down, and I saw him open the door and stand expectantly. Nothing came out though. When I reached him, he was groping about frantically in the empty cab, and giving vent to the finest assorted collection of oaths that ever I listened to."* Holmes seems to like the Growler. *"I will order a four-wheeler"* he tells Doctor Huxtable in one case [10A]; and on another occasion: *Holmes hailed a*

four-wheeler which was passing [10 A].

An interesting comparison can be found in A Study in Scarlet. *"I satisfied myself that it was a cab and not a private carriage by the narrow gauge of the wheels. The ordinary London growler is considerably less wide than a gentleman's brougham."* [10B]

A barouche [10B] is owned by Lady Beatrice Falder of Shoscombe Old Place. *Within a quarter of an hour we saw the big open yellow barouche coming down the long avenue, with two splendid, high-stepping grey carriage horses in the shafts* (Shoscombe Old Place).

In one case (The Man with The Twisted Lip), Holmes uses his fingers between his teeth to give a shrill whistle and a *tall dog-cart* [10B] *dashes up through the gloom.* Holmes pays off the driver and then takes the reins himself for a seven mile drive towards Lee [10B] in Kent. On another occasion Watson records that *an empty dog-cart, the horse cantering, the reins trailing, appeared round the curve of the road and rattled swiftly towards us* (The Solitary Cyclist).

When Holmes and Watson arrive at Leatherhead Station, Watson tells us that: *We hired a trap* [10B] *at the station inn, and drove for four or five miles through the lovely Surrey lanes.* In the same adventure Doctor Rimsby Roylott also rides in a trap (The Speckled Band); and Mr. Scott Eccles travels in one to Esher (Wisteria Lodge). Other stories also mention use of the trap [10B].

Mrs. Tangey takes a time to reach her home because she travels by bus [10B] (The Naval Treaty); and after being kidnapped and then abandoned on Hampstead Heath, Mr. Warren also takes a bus home (The Red Circle).

The same story refers to a newspaper agony column [10B] item about a lady who fainted in the Brixton bus, and the existence of the London bus is acknowledged in another adventure. Lord Mount-James is the miserly but rich uncle of Godfrey Staunton, who goes missing shortly before a rugger match. Presumably to avoid the expense of a cab, Mount-James tells Cyril Overton *"I came round as quickly as the Bayswater bus would bring me"* (The Missing Three-Quarter).

When Watson, Sir Henry Baskerville, and Doctor Mortimer with dog arrive at the railway station in Devon on their way to Baskerville Hall, Watson notes that: *Outside, a wagonette* [10B] *with a pair of cobs* [10B] *was waiting.*

Watson receives an early morning wake-up call one day. *"There's a brougham* [10B] *waiting for us, Watson,"* says Holmes standing by the doctor's bedside (The Resident Patient). On another occasion Watson leaves Baker Street in a brougham driven by Holmes' brother Mycroft (The Final Problem); and we are informed that the famous surgeon Sir Leslie Oakshott also uses a brougham (The Illustrious Client). On the other hand a smart little landau [10B] is owned by Irene Adler (A Scandal in Bohemia); while Colonel Ross has a comfortable one (Silver Blaze).

In one case Holmes nearly gets run over by a two-horse van (The Final Problem). *"As I passed the corner which leads from Bentinck Street on to the Welbeck Street crossing a two-horse van* [10B] *furiously driven whizzed round and was on me like a flash."*

His Last Bow is a first world war spy mystery with several references to cars. Early on in the adventure we learn that Baron von Herling is an important German official from London and that *his huge 100-horse-power Benz* [10C] *car was blocking the country lane as it waited to waft its owner back to London.*

No spy mystery would be complete without a secret code or communication system, and Altamont, who is an Irish-American traitor, sends a telegram to Von Bork, a leading German secret agent, using a simple word code. It states: *Will come without fail to-night and bring new sparking plugs* [10C]. *Altamont.*

Von Bork explains its meaning to Baron von Herling, chief secretary of the German legation in London. *"You see he poses as a motor expert and I keep a full garage. In our code everything likely to come up is named after some spare part. If he talks of a radiator* [10C] *it is a battleship, of an oil pump* [10C] *a cruiser, and so on. Sparking plugs are naval signals."*

The Benz, driven by a chauffeur, allows von Herling to lie back in the cushions of the luxurious limousine. As it proceeds it nearly collides with a small Ford [10C] coming in the opposite direction. This car, also driven by a chauffeur, has Altamont as a passenger. However all is not as it seems, and suffice it to say that we learn later that Watson knows how to drive a car.

The railways were essential to Holmes and Watson, who frequently used the extensive system, as it was then, to reach many parts of the country when called upon to investigate cases outside London. Although they usually travelled in hansoms and horse-drawn vehicles in central London, there was also the underground railway system available for their use. This was gradually being extended throughout the years in which their adventures were taking place. The closest station to the consulting rooms, was of course that of Baker Street. This was served by the Metropolitan Railway [10C], which first opened in 1863 between Bishop's Road (Paddington) and Farringdon Street. This was followed by the Inner Circle line, which started operating in 1884. Later the Baker Street and Waterloo Railway, a deep level tube soon known as the 'Bakerloo' [10C] was constructed, with platforms at Baker Street opening in 1906.

One of the adventures (The Bruce-Partington Plans) has the underground playing a major part in the mystery. It involves the clever use of a train on the Metropolitan Railway to dispose of a corpse. *"The body was found at six on the Tuesday morning. It was lying wide of the metals* [10C] *upon the left hand of the track as one goes eastward, at a point close to the station* [Aldgate], *where the line emerges from the tunnel in which it runs."* At first it is assumed that the person has fallen from a carriage, but Holmes reasons that the lack of a ticket on the body proves that the individual had not entered the system through a station entrance, which would have been protected by a ticket collector. And there is another clue for Holmes as he looks at the track where the corpse has landed. *"Points. And a curve, too. Points, and a curve,"* he mutters, as an idea strikes him.

Later his idea is proven when he and Watson investigate an address that they have obtained. When they ascend to an upstairs room, the sound of a passing train is soon heard. *"Let us stay here until a train stops"* says Holmes. *We had not long to wait. The very next train roared from the tunnel as before, but slowed in the open, and then, with a creaking of brakes, pulled up immediately beneath us. It was not four feet from the window-ledge to the roof of the carriages.* It transpires that after Arthur Cadogan West was murdered, his body was placed on the roof of a stationary train from the window of this particular room, and indeed Holmes finds a bloodstain on the wooden window frame to support this. When the train reached the points and curve near Aldgate station, the jolting motion of the coaches deposited the corpse onto the track.

Another case (The Beryl Coronet) also has an underground railway reference, as does The Red-Headed League. Alexander Holder, a banker in the former case, tells Holmes and Watson when he visits them at 221B: *"I came to Baker Street by the Underground, and hurried from there on foot, for the cabs go slowly through this snow."*
The second case has Holmes and Watson using the underground system to reach the City [10C]. *We travelled by the Underground as far as Aldersgate,* Watson records in his account.

Moriarty hires a special train [10C] to pursue the timetabled Continental express travelling to Newhaven with Holmes and Watson on board, when both are trying to escape his clutches. However Holmes has reasoned out Moriarty's cunning plan, and he and Watson decide to alight at Canterbury ¹⁰ᶜ. While standing on the platform they see that: *far away, from among the Kentish woods there rose a thin spray of smoke. A minute later a carriage and engine could be seen flying along the open curve which leads to the station. We had hardly time to take our place behind a pile of luggage when it passed with a rattle and a roar, beating a blast of hot air into our faces.* (The Final Problem).

After solving the mystery of **Silver Blaze**, Holmes and Watson travel back to London from Winchester ¹⁰ᴰ in a Pullman Car ¹⁰ᴰ. *We had a corner of a Pullman car to ourselves that evening as we whirled back to London.* While Holmes and Watson could occasionally sample some peaceful luxury on a train journey, there was sometimes a more threatening side to railway trips for others. **The Mazarin Stone** is a case that contains a report of a robbery on a Riviera express. Holmes has kept a record of the misdeeds of Count Sylvus in a notebook, and he confronts the Count with a list of some of the entries within it. *"Plenty more here, Count. Here is the robbery in the train-de-luxe to the Riviera* ¹⁰ᴰ *on February 13th, 1892."*

There are numerous cases in the canon which refer to travel on ships or boats. Steam engines are mentioned (**The Sign of Four**) in a fast boat chase on the Thames, as a police launch with Holmes and Watson on board, tries to overtake the *Aurora*, which is speeding away with Jonathan Small and the murderous Andaman islander, Tonga. Doctor Watson describes their progress with gusto: *We were fairly after her now. The furnaces roared, and the powerful engines whizzed and clanked like a great metallic heart.*

The "**Gloria Scott**" adventure and **Black Peter** both have nautical themes. The survivors from the rebellious mutiny of convicts on the *Gloria Scott* are picked up by *the brig* ¹⁰ᴰ *Hotspur, bound for Australia.*

The Cardboard Box also has an important link with the sea, when Holmes tracks down Jim Browner, a ship's steward, who is wanted in connection with the murders of Alec Fairbairn and Mary Browner.

Elsewhere Watson notes in one account *the shocking affair of the Dutch steamship, Friesland, which so nearly cost us our lives* (**The Norwood**

Builder), and **The Abbey Grange** briefly mentions another ship as Captain Croker recounts his story, describing to Holmes how *"was first officer of the Rock of Gibraltar"* when he met Mary Fraser (i.e. Lady Brackenstall before her marriage), who was a passenger on board. In order to track down the murderers of John Openshaw, (**The Five Orange Pips**) Holmes searches through Lloyd's registers [10D] and files for vessels visiting Pondicherry [10D] in January and February 1883. Later, Holmes learns that the perpetrators have left Britain on board *The Lone Star*, an American barque [10D], but he sends them a letter containing five orange pips, observing: *"By the time that their sailing-ship reaches Savannah* [10D] *the mail boat will have carried this letter."* The explanation being that the mail boat (a steamer) is faster than the sailing ship, which however, never arrives in Savannah as it is lost in a storm. Holmes also cables the police in America in the hope that they will make an arrest, but it is all for nothing.

In another adventure Watson takes an evening newspaper to show Holmes, who is in hospital. The paper contains an announcement of relevance to the case on which Holmes is working and simply states that: *Among the passengers on the Cunard* [10D] *boat Ruritania* [10D]*, starting from Liverpool on Friday, was the Baron Adelbert Gruner* (**The Illustrious Client**).

Near the closing of the case in **The Sign of Four**, Holmes is brilliantly loquacious, and at a meal with Watson and Athelney Jones speaks on a variety of subjects including warships of the future. In **The Bruce-Partington Plans** case Holmes is called in to investigate the loss of some important technical papers. These are, we are told, the plans of the Bruce-Partington submarine [10D]. Mycroft, Holmes' brother, comments on the importance of these documents. *"The plans, are exceedingly intricate, comprising some thirty separate patents* [10E]*, each essential to the working of the whole."* **His Last Bow** has the treacherous Altamont telling Von Bork that he has obtained stolen naval signals, referred to in code as *"semaphore* [10E]*, lamp code* [10E] *and Marconi* [10E]*"*.

COMMUNICATIONS

A telegram [10E] was a message that could be telegraphed or dictated over the telephone (see below) for speedy delivery. Specific mention of telegrams appear in several stories [10E] and in The Second Stain one is sent in cipher.

Holmes uses telegraph forms [10E] in Black Peter, The Abbey Grange, and in The Hound of the Baskervilles where at least two are sent. Others use the telegraph system as well. For example the vicar, Mr. Roundhay sends a telegram to Doctor Sterndale in The Devil's Foot; and Holmes receives one from Cyril Overton in The Missing Three-Quarter.

Holmes has a telephone [10F] at his consulting rooms. Use of this instrument by him occurs in three of his later cases [10F]. At the time the instrument was becoming more widespread. However some of his earlier cases do mention it as well. For example the police have telephones (The Sign of Four and The Three Garridebs) at Scotland Yard [10F], and at Bow Street Police Station there is one (The Man with the Twisted Lip). Colonel Emsworth also has an instrument (The Blanched Soldier), perhaps installed about the same time as the one at Baker Street.

Watson on one occasion (The Retired Colourman) finds the instrument useful when he is stranded with Josiah Amberley in a country village. *There was a telephone, however, at the little Railway Arms, and by it I got into touch with Holmes.*

ELECTRICITY

An electricity supply to houses in rural areas was still comparatively rare at the time of the very early stories. However in the now famous Dartmoor story (The Hound of the Baskervilles) of 1901-2, Sir Henry Baskerville intends to wire gloomy Baskerville hall, saying: *"It's enough to scare any man. I'll have a row of electric lamps up here inside of six months, and you won't know it again, with a thousand candlepower*

Swan and Edison [10F] *right here in front of the hall door."*

Of course properties equipped with electricity, whether in town or country, can have electric bells fitted. Holdernesse Hall is wired up with one (The Priory School); and in The Mazarin Stone we find that 221B Baker Street has also had an electric bell installed. It is a good way to summon the police, who are below, when Holmes surprises two villains in his own room. *Before they had recovered Holmes had pressed the electric bell.*

FARMING

John Garrideb is jubilant because he has found a third Garrideb. He arrives at Nathan Garrideb's home waving a newspaper in the air. It contains the advertisement of a constructor of agricultural machinery, based in Aston [10F]. The proprietor's name is Howard Garrideb. *Binders* [10F], *reapers* [10F], *steam and hand ploughs* [10F], *drills* [10F], *harrows* [10F], *farmers' carts, buckboards* [10F] *and all other appliances* are listed in the advert. The company also offers estimates for Artesian Wells [10F] (The Three Garridebs).

MINING

A protection racket is in operation against the local mining companies in the Vermissa Valley (The Valley of Fear), and the Scowrers [10G] make a threat to damage the winding gear [10G] of one company which does not wish fully to co-operate. At the Crow Hill mine, the violence of the Scowrers is witnessed by Scanlan and McMurdo (an undercover agent) when the mine manager is shot in cold blood. This is followed by the killing of Menzies, the mine engineer, who after emerging from the engine-house [10G] sees the first killing and flies at the killers, but is brutally gunned down.

A piercing blast on a steam whistle [10G] is used to call workers to their work, or to alert them to the end of a shift or start of a meal break. The

Crow Hill mine utilizes this system, as the following passage describes: *There came the sudden scream of a steam whistle. It was the ten-minute signal before the cages* [10G] *descended and the day's labour began.* (The Valley of Fear)

HYDRAULICS

There are many references to hydraulics in The Engineer's Thumb, a story which is not really suitable for the squeamish. It is odd how reading an account of a severe injury where the recipient survives, can be more distressing than reading one where the victim is brutally murdered. The engineer and central character in this adventure is Victor Hatherley, who after serving an apprenticeship with the firm of Venner and Matheson of Greenwich, sets up his own chambers (office) in London at 16A Victoria Street [10G]. His specialist engineering skill is in hydraulics [10G]. Hatherley is engaged by Colonel Lysander Stark to look at a hydraulic stamping machine which has *got out of gear* [10G]. He travels to Berkshire and is told that the hydraulic press [10H] he has come to see, is used in the processing of fuller's earth [10H]. The Colonel and Hatherley enter a small room which is the actual hydraulic press itself, and as the Colonel remarks:

"it would be a particularly unpleasant thing for us if anyone were to turn it on. The ceiling of this small chamber is really the end of the descending piston, and it comes down with the force of many tons upon this metal floor. There are small lateral columns of water outside which receive the force, and which transmit and multiply it in the manner which is familiar to you. The machine has lost a little of its force."

After examining the machinery, Hatherley finds the problem is caused by a slight leakage which allows a regurgitation of water through one of the side cylinders [10H]. He discovers that a rubber band [10H] round a driving-head rod has shrunk and is causing the loss of power, as he points out to the Colonel and his colleague Ferguson, who is described as the secretary and manager. An excellent adventure story cliff-hanger ensues as tension builds up. Hatherley is trapped under the press as it slowly descends upon him. Is escape possible? At the end of the adventure, after fire destroys the evidence of the existence of the press, all that is left are some twisted cylinders and iron piping.

Holmes later brings an old news cutting to the attention of Hatherley, following his remarkable escape from the counterfeiting gang who were operating the press. This cutting reports the disappearance of another hydraulic engineer, Jeremiah Hayling, about a year before Hatherley's ordeal. They both sadly conclude that Hayling was not so lucky, and his fate was sealed by the same gang.

HANDCUFFS

Several stories report that handcuffs or 'darbies' [10H] are clapped on miscreants when they are caught. Two examples are given here: *"Stand still, will you?" There was the click of the closing hand-cuffs* [10H]. This is the scene when Culverton Smith is caught by the cunning brilliance of Holmes, and Inspector Morton of the official police has the honour of putting the bracelets on Smith (The Dying Detective).

In another case (Black Peter) Watson recounts what happens when Holmes fastens the handcuffs onto the murderer Patrick Cairns. *I heard a click of steel and a bellow like an enraged bull. The next instant Holmes and the seaman were rolling on the ground together.*

Implements put to a nefarious use are mentioned in some stories. Thus in A Scandal in Bohemia Holmes instructs Watson to throw a smoke-rocket [10H] and create a diversion at the home of Irene Adler, after Holmes has been carried inside. *"It is nothing very formidable"* he said, taking a long cigar-shaped roll from his pocket. *"It is an ordinary plumber's smoke-rocket, fitted with a cap at either end to make it self-lighting."*

On another occasion when Watson is alone at the Baker Street consulting rooms, he receives a note from Holmes via a messenger. The message asks him to join Holmes at a restaurant in Kensington. He urges: *Bring with you a jemmy* [10H], *a dark lantern* [10H], *a chisel, and a revolver. There is obviously unlawful work to be done!* (The Bruce Partington Plans).

HOME COMFORTS

A gasogene [10I] in the Baker Street rooms is referred to in two of the stories. In The Mazarin Stone Holmes remarks that: *"The gasogene and cigars are in the old place."* In another story (A Scandal in Bohemia) Holmes points to the *spirit case* [10I] *and a gasogene in the corner*, and so invites Watson to use them. In Black Peter Peter Carey has *"a tantalus* [10I] *containing brandy and whisky on the sea-chest."*

It is inevitable that Holmes as a music lover would purchase a machine on which to play gramophone records. *"These modern gramophones* [10I] *are a remarkable invention,"* he observes, after using it to play a record of a violin piece. In doing so, he manages to deceive the two villains left on their own in his consulting room, into thinking that he is actually playing the piece himself in the next door room. Meanwhile he has overheard some very interesting information concerning a precious stone (The Mazarin Stone).

Since the publication of the last Sherlock Holmes story in 1927, there has been an unstoppable advance of technology. Computers, mobile phones, and satellite tracking devices are some of the many machines used in the ongoing fight against crime. The development of more reliable and faster transport has revolutionized the street and rail scene from Holmes' day, and he would have approved of all these changes, because his keen mind was always ready to adapt to new ideas and progress, as exemplified by his own approach to research.

NOTES 10A

The bicycle was invented in 1839 by Kirkpatrick Macmillan, a Scottish blacksmith. References to the bicycle are found in: The Priory School, The Solitary Cyclist, The Valley of Fear and The Missing Three-Quarter.

Dunlop tyres: The name originated from John Boyd Dunlop (1840-1921), a Scotsman by birth who settled in Ireland, and invented the first pneumatic rubber bicycle tyre in 1888.

Tyre tread: Some writers have said that this deduction could not be valid, as the tracks would look the same whichever way the cyclist had travelled. However, depending on the actual pattern of the tyre tread and if this was known beforehand, such a deduction could possibly be made.

Unbreakable tyre: Some bicycles had solid rubber tyres.

The Hansom was a favourite form of transport for better-off Victorians and Edwardians. A horse-drawn precursor of the modern London taxi cab, but with the driver perched high at the back of the vehicle (although the early version had the driver at the front). There was a single horse, one very large wheel each side, with the passenger entering via a platform and then through a pair of padded half-doors at the front of the cab. It was designed in 1834 by the architect Joseph Aloysius Hansom, from whom it obtained its name, but was later greatly improved by John Chapman who patented his design in 1836.

Reference to transport by hansom is found in: The Final Problem, The Veiled Lodger, The Red-Headed League and others.

The Growler, or Clarence cab, was a four wheeled horse-drawn vehicle in use at the time. The popular design of this cab dates from the 1830s (though some four-wheeled cabs were in use in the 1820s), and required one or two horses to pull the vehicle. It usually carried four passengers. The name Growler was a London slang term given to the vehicles.

References to the Growler are found in The Priory School and The Blue Carbuncle.

NOTES 10B

Barouche: A horse-drawn four-wheeled carriage with two seats facing each other. It was spacious and had an adjustable hood.

Dog-cart: Usually a two-wheeled vehicle drawn by a single horse. It could seat two people, with dogs travelling under the seat. **The Trap** is a similar light two-wheeled carriage. It is mentioned in The Solitary Cyclist and The Man with the Twisted Lip.

Lee: Now a suburb of South-East London.

Bus or omnibus: George Shillibeer introduced the horse bus to London in 1829 and called it an 'Omnibus' (meaning 'for all'). When other operators started horse-bus services to cash in on his publicity, some put 'shillibeer' with a small 's' on the sides of their vehicles, and a tiny 'not' in front of the word. Clearly a case of mind the small print! When the early Sherlock Holmes adventures were in vogue, the horse buses in London required two horses to pull them, and were of two different types, based on the seating layout of the vehicles. These were double-decked and had open-tops and acquired the names of 'knifeboard' or 'garden seat' buses.

Agony column: A newspaper column for personal advertisements.

Wagonette: A four-wheeled horse-drawn carriage with inward-facing seats and rear access. Ideal for conveying passengers with luggage or picnic hampers.

Cobs: Short-legged, sturdy horses, ideal for driving or riding as long as stamina rather than speed is required.

The Brougham was a private four-wheeled carriage named after Lord Brougham who introduced it in 1838. The advantage of it was its lightness so that it only required one horse. There were different models and some were given nicknames. The 'pillbox' was said to be the doctors' favourite!

Landau: A four-wheeled carriage usually drawn by two horses. The coach is suitable for use in both winter and summer as it has an adjustable hood, which can be opened, half-opened or closed.

Two-horse van: Probably a carrier's delivery vehicle.

NOTES 10C

Benz motor car: This was a German car named after Karl Benz, who designed his own vehicle in about 1885. (1 horse power = the power to lift 550 pounds by one foot in a second.)

Sparking plugs: An essential part of the engine, attached to the electrical system, providing the initial spark to ignite the fuel supplying the car engine.

Radiator: At the front of the car, a water-cooled metal grille to dissipate heat from the engine.

Oil pump: Operated by the engine crankshaft, and supplies lubricating oil from the oil tank to the bearings or working parts of the engine.

Ford motor car: Named after Henry Ford, who was the pioneer of mass production for cars.

Metropolitan Railway: Then an independent railway company, now the Metropolitan line on the London underground system.

Bakerloo: Captain G.M.F. Nichols, 'Quex' of the *Evening News*, coined the word 'Bakerloo'.

Metals: The rails on which the trains ran.

City with a capital 'C' refers specifically to the 'City of London' or 'Square Mile', the small area of London containing many prominent financial institutions.

Special train: Engine hauled saloon coaches were available for hire on the railways, but could Moriarty have hired one at such short notice?

Canterbury: An important cathedral city in Kent, south east of London.

NOTES 10D

Winchester: A cathedral city in the county of Hampshire, south west of London.

Pullman Car: These luxuriously designed railway carriages had superior interiors and were first introduced into British service in 1874 by the American, George Mortimer Pullman. The London and South Western Railway used them on its Bournemouth to London service via Winchester.

The Riviera: In this context, the South of France, Monaco and North West Italy.

Brig: Shortened name for brigantine, a sailing ship with two masts.

Lloyd's Register: A society with origins dating from 1760 which still exists today. Its purpose was to classify and survey merchant ships, and to maintain and publish an annual register of them.

Pondicherry: A town in India.

American barque: A sailing ship, often three-masted.

Savannah: A city in Georgia, U.S.A.

Cunard Line: Shipping company which operated steamships and was founded by Sir Samuel Cunard in 1839, with the first voyage in 1840.

The *Ruritania*: Almost certainly a fictitious name, perhaps derived from the name of the kingdom in the Anthony Hope novel *The Prisoner of Zenda* (1894).

Submarines have a long history dating from the 17th century, but the 19th century was the period of greatest development for this type of vessel as mechanically propelled engines became available to drive them.

NOTES 10E

Patents: To protect new inventions from unfair exploitation and from being copied, patent specifications are drawn up accurately to describe and illustrate the design and working of the invention. Protection only lasts for a limited number of years, and patents have to be taken out in different countries to avoid widespread infringement.

Semaphore: A hand operated signalling device used by sailors to communicate with other ships while at sea. Two movable arms at various positions at the top of a post were used to transmit letters of the alphabet.

Lamp code: A coded system using lamps at night or in poor visibility to signal between ships.

Marconi: Guglielmo Marconi (1874-1937) was an Italian physicist who invented radio, a technology which has proved to be of the utmost value to ships for communication and navigation purposes.

The electric telegraph: The first efficient telegraph communication system was set up in 1816 (with details published in 1823) by Sir Francis Ronalds of Hammersmith, London. Messages could be sent in plain language or in code (such as the Morse system) using a finger-operated keypad. Pre-punched paper tape could operate the transmitting apparatus at the sending end, and paper tape machines could print out messages at the receiving end.

Telegrams: The telegraph system was used to send telegrams to individuals. The sender wrote a message on a form, the cost depending on the length of the message, which was then telegraphed through to the nearest telegraph office or main post office, from which telegrams were dispatched. At the receiving office the message had to be transcribed into plain English if relayed using code, and then transferred to the paper telegram for fast onward delivery by courier to the recipient's address.
References to telegrams are made in The Devils Foot, The Reigate Puzzle and A Study in Scarlet.

NOTES 10F

The telephone was invented by the Scotsman, Alexander Graham Bell (1847-1922) in 1875 and was patented the following year. References to the telephone are found in: The Illustrious Client; The Retired Colourman; The Three Garridebs. These were published in the 1920s.

Scotland Yard. The headquarters of the Metropolitan Police near the Embankment in London.

Swan and Edison lamps were made by the Edison and Swan United Electric Light Company, formed by the merger in 1883 of the two great pioneering electricity companies, namely the Edison Company founded by the American inventor Thomas Alva Edison, and the Swan Company founded by the British inventor, [Sir] Joseph Wilson Swan.

Aston: A district of Birmingham in the Midlands region of England.

Binders: Machines that tied the sheaves or bales.

Reapers: Crop cutting machines.

Steam and hand ploughs: These excavated furrows to prepare the ground for sowing.

Drills: Machines which sowed seeds in rows.

Harrows: Machines which filled in the furrows, broke up lumps of earth or removed weeds from ploughed land.

Buckboard: A flat-backed wooden cart for general use. An American version had seats.

Artesian wells: Bored ground wells to reach underground watercourses. Hydrostatic water pressure enables the water to rise to ground level of its own accord. The name is derived from Artois, the district around Arras in France, where the earliest well dating from the 12th century was sunk.

NOTES 10G

Engine-house: The building housing the engine driving the winding gear (or winch) of the mine.

The Scowrers: The Eminent (or Ancient) Order of Freemen is a fraternal secret society, which is fictitious, but probably based on 'The Molly Maguires'. This was a real Irish/American secret society for Pennsylvanian miners, in operation before 1880.

Winding gear: The mine machinery consisting of wheels and belts which raise and lower the cages for the workers and bring the product and spoil to the surface. If this was damaged, the mine could not operate.

Steam whistle: A valve operated this whistle, which would have been connected to the steam boiler providing the power for the winding gear. Intended as a safety device, it first appeared in 1826.

Cages: The lifts that took the miners up and down to the excavation working area of the mine.

Victoria Street: Anyone who knows London well today, may be aware that the important professional bodies for engineers are situated in the Victoria Street area of the city. It is natural then that an engineer would set up his office near his professional Institution.

Hydraulic engineering is that branch of technology that deals with plant and machinery concerned with, or affected by the properties and the flow of fluids.

"Got out of gear": Not synchronized or working properly.

NOTES 10H

A hydraulic press operates by using water pressure to create and magnify a force, which is then utilized in the pressing operation. One of the earliest successful hydraulic presses was invented by the engineer, Joseph Bramah, who obtained a patent for it in 1795.

Fuller's earth is a clay of some fineness, with good absorbent properties. It was used for fulling cloth, hence the name, and in a wide range of other industrial processes. There are two types of fuller's earth, the one mentioned here is almost certainly the variety that contains montmorillonite.

Side cylinders: The pistons of the press operate in these cylinders.

Rubber: Derived from trees and plants as a liquid rubber latex which can solidify, be dried and treated to produce raw rubber.

Darbies or Derbies: The slang term for handcuffs is used in The Cardboard Box and The Red-Headed League. The origin of the word is unclear.

Handcuffs: Almost certainly developed from metal manacles used for securing prisoners.

A smoke-rocket was used by plumbers to determine either the course of a drain and/or whether it was secure from leaks. If there was a leak, the smoke would seep out at the relevant place, but it only worked satisfactorily for short lengths of pipe. The cap at the end of the rocket was a tiny fuse which could be activated in order to fire it.

Jemmy: A metal crowbar used by burglars to break into premises.

Dark lantern: It seems a contradiction in terms, but it is a lantern which has a shutter or hinged plate to cover or reduce the light.

NOTES 10I

A **gasogene** was an early form of soda water siphon and sometimes called a Seltzogene. It consisted of two glass globes one on top of the other, with the lower larger than the upper (see illustration), and connected by glass tubes. At the top of the apparatus there was a trigger which operated a valve, and also a short exit tube for the carbonized liquid. The soda water was made by placing sodium bicarbonate and, usually, tartaric acid in the upper globe and water in the lower one. After fastening the top, the apparatus was tilted to mix the powders and water. This created carbon dioxide which, dissolved in the water, could then be released under pressure as required, using the valve. The glass container was usually enclosed in a wire mesh in case the pressure inside it caused it to shatter.

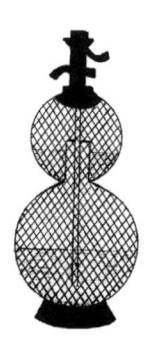

Spirit case: A case to hold glass bottles or decanters containing spirits.

A tantalus was a set of glass decanters for spirits, with the tops secured to prevent unauthorized persons, especially servants, from gaining access to the contents of the decanters.

The gramophone developed from the phonograph. It was a mechanical machine which was spring-operated, required winding, and played recorded music on either wax cylinders or discs of shellac (made from a resin produced by the lac insect).

11 ANIMAL OR VEGETABLE ?

NATURAL HISTORY

The amount of detail in the Sherlock Holmes adventures and novels is astounding. Conan Doyle, as well as being a good observer, clearly had a wide knowledge of a whole range of subjects, not least the many plants and animals that are mentioned. They often appear in short descriptive passages, but sometimes have a major part to play in the cases that Holmes is investigating such as The Speckled Band and The Lion's Mane.

THE ANIMAL KINGDOM

In an instant his strange headgear began to move (The Speckled Band).

The Speckled Band, which Holmes investigates, has a number of unusual creatures in it, but the speckled band of the title is a snake – a deadly Indian swamp adder [11A]. Doctor Roylott uses this snake to commit murder on his behalf. However one night when Holmes and Watson are keeping watch, Holmes spots the creature and lashes out at it with his cane. The snake returns to Roylott's room where it is kept, but turns on its owner. Watson's account describes the moment: *There broke from the silence of the night the most horrible cry to which I have ever listened. It swelled up louder and louder, a hoarse yell of pain and fear and anger all mingled in the one dreadful shriek.* When Holmes enters the room, he finds Doctor

Roylott dead and 'wearing' a strange brown and speckled yellow band around his head. *"The band! the speckled band!" whispered Holmes.* As Watson steps forward, the snake moves and they both see *the squat diamond-shaped head and puffed neck of a loathsome serpent.*

Vipers [11A] and a venomous lizard or gila [11A] appear in the 'V' index volume of Holmes' past cases, which Watson retrieves from the shelf at their Baker Street rooms. (The Sussex Vampire).

Mr Sherman, the South London taxidermist visited by Watson, keeps at his premises a slow-worm [11A] which eats beetles [11A]. In the same adventure, another reptile – a crocodile is responsible for Jonathan Small having only one good leg after swimming in the Ganges [11A]. *"A crocodile took me, just as I was half-way across, and nipped off my right leg as clean as a surgeon could have done it, just above the knee."* (The Sign of Four).

Also to be mentioned here is the snake-catcher or ichneumon [11A]. In The Crooked Man Henry Wood tells Holmes that his pet snake-catcher, Teddy *"is amazing quick on cobras* [11A]. *I have one here without the fangs, and Teddy catches it every night to please the folk in the canteen."* It must have really put the diners off their meal!

Worms are seen to have eaten their way through the wooden box Holmes finds in the chamber beneath the cellar of Hurlstone. (The Musgrave Ritual).

In The Problem of Thor Bridge Watson tells us that in the vaults of a London bank, he has left a dispatch-box which contains many reports of the cases of Sherlock Holmes. Amongst these are some that were complete failures. These include *that of Isadora Persano, the well-known journalist and duellist, who was found stark staring mad with a match box in front of him which contained a remarkable worm, said to be unknown to science.* Another case, though if it failed we are not told, is referred to in The Golden Pince-Nez. This was *the repulsive story of the red leech* [11B] *and Crosby the banker.*

Some writers of pastiche stories about Holmes have used so-called 'failed' cases and others mentioned in passing in the canon, as storylines on which to base their own Sherlock Holmes adventures. [11B]

The Creeping Man is a story about a rejuvenation elixir and its effect. The serum that Professor Presbury takes, comes from the black-faced langur monkey [11B] (which can crawl and climb). Lowenstein is the scientist who supplies the product, but he suggests that a serum from an anthropoid [11B] would be more suitable, as such a creature walks with an upright stance.

Doctor Roylott is fascinated by Indian animals and keeps a cheetah [11B] and baboon [11B], which are allowed to roam freely in the grounds of his house. As Holmes and Watson cross the lawn they are startled by *what seemed to be a hideous and distorted child, who threw itself on the grass with writhing limbs, and then ran swiftly across the lawn into the darkness.* Holmes realizes that this must have been the baboon. And while they mount vigil in Miss Stoner's room one night, they hear *a long drawn, cat-like whine, which told us that the cheetah was indeed at liberty.* (The Speckled Band).

Two more wild animals, this time native to America, appear in A Study in Scarlet. *The coyote* [11B] *skulks among the scrub, and the clumsy grizzly bear* [11C] *lumbers through the dark ravines.* Smaller animals are found in another adventure (The Sign of Four), in which, following the murder at Pondicherry Lodge, Holmes directs Watson to a bird-stuffer named Sherman in Lambeth, telling him that *"You will see a weasel* [11C] *holding a young rabbit in the window."* When Watson arrives, Sherman warns him to *"Keep clear of the badger, for he bites."* There is also a stoat [11C] that thrusts *its wicked head and red eyes between the bars of its cage.*

An animal of *"the weasel and stoat tribe"* is suspected of playing a part in The Crooked Man mystery. Holmes deduces that the animal, although he has not seen it, has *"a long body with very short legs attached to it, can run up a curtain, and is carnivorous."* Further into the adventure, the reader finds out that Henry Wood is the owner of this creature and that he carries it about with him in a box. His landlady though is *"in considerable trepidation, for she had never seen an animal like it"*. As already mentioned it is eventually revealed that the mysterious animal is a mongoose [11A] or ichneumon [11A], named Teddy.

The Lion's Mane case is based on the harmful stinging jellyfish *Cyanea capillata* [11C]. Holmes finds a description of its painful effect on humans in a small book he owns: *The multitudinous threads caused light scarlet lines upon the skin which on closer examination resolved into minute dots or pustules, each dot charged as it were with a red-hot needle making its way through the nerves.*

One story which has several animal and plant references is The Hound of the Baskervilles. The plot is in a league of its own, and it is brimming with natural history content, much of it relating to lonely Dartmoor. It also has an important character who is a naturalist. This is Cyril Stapleton, whose hunting ground is the desolate moor which he appears to know intimately. He collects butterflies, as well as botanical specimens that he places in a tin box [11J] carried over his shoulder.

Mr. Sherman is also a naturalist, but in a very different way, as he specializes in taxidermy [11C], although he also keeps live animals in his home in Pinchin Lane, Lambeth. It is here that Doctor Watson on one occasion visits him. In the following excerpt Watson describes Sherman's approach: *He moved slowly forward with his candle among the queer animal family which he had gathered round him. In the uncertain, shadowy light I could see dimly that there were glancing, glimmering eyes peeping down at us from every cranny and corner* (The Sign of Four). Other characters in the adventures who have some connection, usually damaging to natural history include Doctor Sterndale, *the great lion-hunter and explorer (*The Devil's Foot*);* and

Edward, the Rucastles' son, an unpleasant child who plans *"the capture of mice, little birds and insects"*. (The Copper Beeches). He also liked killing cockroaches [11C] with a slipper, but can we condemn him for that, as long as he cleaned the slipper afterwards?

In The Five Orange Pips Holmes muses on the philosophy and science of deduction by comparing it to the skill of Cuvier [11C], who he said could describe *"a whole animal by the contemplation of a single bone"*.

The case of The Priory School brings to light the clever subterfuge of replacing common horseshoes by special iron ones shaped like cloven feet. The Duke of Holdernesse informs Holmes that reputably they belonged to the Barons of Holdernesse in the Middle Ages, and were used to throw off-track any pursuers who tried to follow them.

Nathan Garrideb keeps his own small but fascinating museum of geological and natural history exhibits. As described in the story (The Three Garridebs), this room *was both broad and deep, with cupboards and cabinets all round, crowded with specimens, geological and anatomical.*

A reference to a giant rat concerns one of Holmes' past cases about the *Matilda Briggs*, a ship which was *"associated with the giant rat of Sumatra*[11D]*, a story for which the world is not yet prepared"* as Holmes tells us in The Sussex Vampire.

Moles[11D] can be a pest to keen gardeners, and Doctor Watson delights in an exaggeration in the following quote: *Holmes peered at the great rubbish-heaps which cumbered the grounds. "It looks as though all the moles in England have been let loose in it"*, says Watson. (The Sign of Four)

BIRDS

One day, disguised as a bookseller, Holmes calls at Watson's Kensington home, and offers to sell him some books: *"Maybe you collect yourself, sir. Here's 'British Birds'..."* (The Empty House).

The Hound of the Baskervilles mentions a number of birds. Stapleton tries to explain away the cry of the hound to Watson by suggesting that it is really a bittern [11D] making the noise. Watson then sees a pair of croaking ravens [11D] on Dartmoor, and later when alone on the moor, he observes from a distance *one great grey bird, a gull* [11D] *or a curlew* [11D], *soar aloft in the blue heaven.*

Curlews also feature (The Priory School), along with plovers [11D], on the stretch of lonely moorland known as Lower Gill Moor; and buzzards appear in a description of the desert in A Study in Scarlet. These large, heavy flapping buzzards [11D] approach and gather to watch the exhausted and parched John Ferrier and the child named Lucy: *Three large brown birds circled over the heads of the two wanderers, and then settled upon some rocks which overlooked them. They were buzzards, the vultures of the West, whose coming is the fore-runner of death.* Conan Doyle's poignant description says it all, but when rescue comes in the form of the Mormons [11E], the three buzzards, cheated of their bounty, utter raucous screams of disappointment and flap sullenly away.

A search of the pond in the grounds of the Abbey Grange reveals some stolen silver, but it is also seen to be the haunt of a familiar majestic bird. *It was frozen over, but a single hole was left for the convenience of a solitary swan* (The Abbey Grange).

Holmes enjoys a woodcock [11E] for his meal at the satisfactory conclusion of his investigation of The Blue Carbuncle; and

Mr. and Mrs. Moulton are invited to a supper which includes a brace of woodcock in The Noble Bachelor.

A game-cock [11E] is on the inn sign outside the Fighting Cock Inn (The Priory School); and a voodoo [11E] ceremony requiring the slaughter of a white cock [11E] by a mulatto [11E] is referred to in Wisteria Lodge.

"You do occasionally find a carrion crow among the eagles", declares Holmes, when explaining the presence of a bad person in an old established aristocratic family. The comparison is plain to see. The crow [11E] is dark and evil, or so it can seem, and lives on carrion [11E]; whereas the eagle is noble and impressive, a 'prince' of birds. (Shoscombe Old Place).

DOGS

Many dogs feature in the Sherlock Holmes stories, but The Hound of the Baskervilles is the adventure with the most references. The curse of the Baskervilles originates from an old story of a pack of hounds that is set upon a yeoman's daughter, who is fleeing from the unwanted attentions of Hugo Baskerville. The terrifying hound itself is inevitably mentioned several times in the narrative, and details of when and where it was purchased are only discovered by the reader towards the end of the novel. At this point it is revealed that the dog was bought in the Fulham Road[11E] in London from the firm of Ross and Mangles. Holmes tells Watson: *"It was the strongest and most savage in their possession."* Considering the ferocity of the brute, the name of 'Mangles' for the dealer is remarkably apt. Before this point in the story, however, there are some good accounts of the hound's mournful and frightening howl, and of course the gripping account of how, with its face covered in luminous paste [11E], it chases Sir Henry Baskerville over the moor. *With long bounds the huge black creature was leaping down the track, following hard upon the footsteps of our friend.* Holmes and Watson, who are waiting and watching, are paralysed, perhaps not with fear, but almost certainly with shock, and let the creature pass before they recover their equilibrium and both fire simultaneously at the ferocious beast. *We saw Sir Henry looking back, his face white in the moonlight, his hands raised in horror, glaring helplessly at the frightful thing which was hunting him down.*

The other dog to play a smaller part in the story, is Doctor Mortimer's curly-haired spaniel [11F], which disappears on the moor. Before Holmes even meets Doctor Mortimer, he deduces that the man has a favourite dog. He does this by scrutinising a stick left behind by the Doctor in his consulting rooms, telling Watson that the dog *"has been in the habit of carrying this stick behind his master. Being a heavy stick the dog has held it tightly by the middle, and the marks of his teeth are very plainly visible. The dog's jaw, as shown in the space between these marks, is too broad in my opinion for a terrier* [11F] *and not broad enough for a mastiff* [11F]*."* Much later, the skeleton of the spaniel *with a tangle of brown hair adhering* is found inside the lair of the hound. Holmes draws the sad and only conclusion possible.

Another spaniel, this time a black one and an example of *the real Shoscombe breed* [11F], plays an important part in solving the mystery which centres on the house called Shoscombe Old Place (also the title of the story). A dog will usually recognize its owner, and is very likely to give away any stranger who tries to deceive it, as occurs in this case. After carrying out a little experiment with the spaniel, Holmes makes the remark that *"Dogs don't make mistakes"*.

The other adventure where a dog proves his worth is The Sign of Four. Toby we are told is *a queer mongrel with a most amazing power of scent*. Watson has to

168

fetch him from Mr. Sherman, a taxidermist in Lambeth, and take him back to Bartholomew Sholto's house, from where both he and Holmes follow the scent of a creosote [11F] trail across South London, using Toby's superior nose to sniff it out. Note the clever simile in the following quote: [Holmes] *pushed the creosote handkerchief under the dog's nose, while the creature stood with its fluffy legs separated, and with a most comical cock to its head, like a connoisseur sniffing the bouquet of a famous vintage.*

Other dogs appear here and there in the adventures. Doctor Watson tells Holmes that he keeps a bull pup [11F](A Study in Scarlet); Holmes is bitten on the ankle by Victor Trevor's bull terrier [11F] (The "Gloria Scott"); and a gold pin with a bulldog's head is found by Gregson, the Scotland Yard detective. (A Study in Scarlet).

"Pompey is the pride of the local draghounds [11F] *– no very great flier, as his build will show, but a staunch hound on a scent."* (The Missing Three-Quarter). Here Holmes makes his comments on the prowess of a sporting dog bearing the Roman name of Pompey. The detective has hired or borrowed the hound to follow the carriage of the mysterious Doctor Armstrong. Holmes has sprayed one of the carriage wheels with aniseed [11F], thus providing an excellent scent for Pompey to follow.

In The Creeping Man Holmes tells Watson that he is thinking *"of writing a small monograph upon the uses of dogs in the work of the detective. A dog reflects the family life. Whoever saw a frisky dog in a gloomy family, or a sad dog in a happy one? Snarling people have snarling dogs, dangerous people have dangerous ones."* Watson is unconvinced of the merits of such a work, but it turns out that Holmes is really considering a more specific problem, which is: *"Why does Professor Presbury's wolfhound* [11F]*, Roy, endeavour to bite him?"* In making these comments, it is surprising that Holmes does not come out with the old chestnut that some people look like their dogs; or some dogs look like their owners.

Another fierce dog, Mr Rucastle's mastiff [11F] Carlo, causes alarm in the household at the Copper Beeches and is described in the words of Miss Hunter as *"a giant dog, as large as a calf, tawny-tinted, with hanging jowl, black muzzle, and huge projecting bones."* (The Copper Beeches)

In A Study in Scarlet Holmes asks Watson to bring upstairs the landlady's terrier, which is suffering badly and close to death. He intends to try out a pill, which he thinks is poisonous, on the dog, and so put it out of its misery.

The Reigate Puzzle involves an investigation of a burglary by Holmes. In the course of his inquiries he asks Mr. Cunningham: *"You don't keep a dog?"* To which Cunningham replies: *"Yes, but he is chained on the other side of the house."*

In The Lion's Mane Holmes is surprised to learn that Mr. McPherson's dog, an Airedale terrier [11G], has *"died of grief for its master"* at the very place where McPherson's body was found on the beach. Holmes' mind starts working overtime: *Why should this lonely beach be fatal to it?* he ponders. It turns out to be one more clue in the investigation of this mysterious death.

CATS

While dog references are numerous in the stories, there are fewer excerpts relating to the domestic cat. Some are merely expressions or comparisons, but below are a few referring to the real animal.

One appears in The Norwood Builder when Mrs. McFarlane tells Holmes of the cruelty (though to birds) of Jonas Oldacre. *"I heard a shocking story of how he had turned a cat loose in an aviary."*

Three further examples follow. In one (Charles Augustus Milverton) Watson writes that: *Something rushed out at us, and my heart sprang into my mouth, but I could have laughed when I realized that it was the cat.*

Martha, the servant of the German spy Von Bork, is observed bending over her knitting and stopping occasionally to stroke a large black cat upon a stool beside her. (His Last Bow)

A cottage holds the mystery at the centre of another case (The Yellow Face) that Holmes investigates. Grant Munro, the detective's client, reveals to him that when he gained access and rushed into the building, *"in the kitchen a kettle was singing on the fire, and a large black cat lay coiled up in a basket."*

HORSES

With the exception of The Silver Blaze adventure, which is wholly concerned with a missing racehorse, horses only play a minor role in the stories. However horse-drawn vehicles such as hansoms, growlers, broughams and landaus are frequently used by characters throughout the adventures. Also two horses, or more accurately a pair of cobs [11G], haul the wagonette [11G] in which Sir Henry Baskerville, Watson and Doctor Mortimer travel; and in the same story a moorland pony is swallowed up by the mire on the moor. (The Hound of the Baskervilles).

A mustang [11G] called Poncho is owned by Lucy Ferrier in A Study in Scarlet, and in the same story she rides on a mule [11G]. Two splendid, high-stepping grey carriage horses are in the shafts of the yellow barouche conveying the impostor in Shoscombe Old Place; and another pair of greys pulls the brougham of Doctor Armstrong in The Missing Three-Quarter.

In one mystery (A Scandal in Bohemia), the horses are valued by Holmes when he glances from the window of his consulting room and sees below *"A nice little brougham and a pair of beauties. A hundred and fifty guineas* [11G] *apiece. There's money in this case, Watson, if there is nothing else."* The reader can almost hear him rubbing his hands together.

INSECTS

In The Hound of the Baskervilles Holmes describes Stapleton as devoted to entomology (the study of insects) and finds out from the British Museum [11G] that he was a recognized authority upon the subject. When Stapleton with his wife moved south, they changed their name from Vandeleur, which was the name given to a moth that Stapleton had been the first to describe when in Yorkshire. The story has several references to Stapleton and his interest in butterfly or *Lepidoptera* collecting (which includes moths) [11G]. When Watson first meets him, he is carrying a green butterfly net, and later he notices it again from a distance, when he observes Stapleton rapidly approaching Sir Henry Baskerville and 'Miss' Stapleton: *He was running wildly towards*

them, his absurd net dangling behind him. It seems that Watson had little or no time for butterflies. Earlier, Stapleton had invited him to *"come upstairs, Dr. Watson, and inspect my collection of Lepidoptera. I think it is the most complete one in the south-west of England."* But Watson, who is keen to return to Baskerville Hall, declines; and also Stapleton's kind offer of lunch.

Grimpen Mire is considered by Stapleton to be the best place for butterflies, and on one occasion as Watson looks on, he chases with his net a specimen of *Cyclopides* [11H] that is flying towards it. Holmes uses an interesting and clever 'butterfly' analogy for Stapleton on realising he is a relation of the Baskervilles. *"We have him, Watson, we have him, and I dare swear that before tomorrow night he will be fluttering in our net as helpless as one of his own butterflies. A pin, a cork, and a card* [11G]*, and we add him to the Baker Street collection!"* Finally on solving the mystery, Holmes and Watson have to rescue Mrs. Stapleton (as she turns out to be) from imprisonment in the room containing Stapleton's insect collection: *The room had been fashioned into a small museum, and the walls were lined by a number of glass-topped cases full of that collection of butterflies and moths, the formation of which had been the relaxation of this complex and dangerous man.*

Another man with a passion for collecting is found in **The Three Garridebs**. Watson in his written account describes Nathan Garrideb's home museum with its cases of butterflies and moths flanking each side of the entrance.

It is revealed (His **Last Bow**) that Holmes in retirement becomes a beekeeper [11H] in Sussex and writes a book with the title: *Practical Handbook of Bee Culture, with Some Observations upon the Segregation of the Queen.* He then uses this book as a substitute for a code book to trap Von Bork, a German spy.

Holmes seems to delight in synonyms and metaphors, and in the following quote from the same adventure, compares gangs of bees to gangs of criminals. *"Behold the fruit of pensive nights and laborious days when I watched the little working gangs as once I watched the criminal world of London."*

In another story (The Dying Detective), beeswax [11H] is used by Holmes to encrust his lips as part of his deception of feigning illness in order to trap the repulsive Culverton Smith.

Jonathan Small is bitten by mosquitoes [11H] whilst a prisoner in the Andaman Islands [11H]. He also refers to the old quarter of the fort at Agra [11H] *"which is given over to the scorpions* [11H] *and the centipedes* [11H]*."* (The Sign of Four)

THE VEGETABLE KINGDOM

Sherlock Holmes is in the guise of a bookseller when Watson bumps into him one day in the street, causing him to drop several books. Watson later records: *I remember that as I picked them up, I observed the title of one of them, 'The Origin of Tree Worship', and it struck me that the fellow must be some poor bibliophile* [11H] (The Empty House).

An oak and an elm tree in the grounds of Hurlstone are important fixed reference points for the working out of the mystery of **The Musgrave Ritual.** *Right in front of the house, upon the left-hand side of the drive, there stood a patriarch among oaks* [11I]. *"It was there at the Norman Conquest in all probability. It has a girth of twenty-three feet,"* as Musgrave informs Holmes. But the elm [11I] has since gone. *"There used to be a very old one over yonder, but it was struck by lightning ten years ago, and we cut down the stump."* The scar of the place where it stood can be seen on the lawn, and Holmes is surprised when Musgrave is able to give him its exact height of sixty-four feet, a measurement that he learnt in his youth. This enables Holmes to measure and make the necessary calculations essential for finding a long lost treasure with Royal connections.

From the station, the road to Baskerville Hall passes through a valley dense with scrub oak [11I] and fir [11I]; and the alley at Baskerville Hall where Sir Charles Baskerville is found dead, is bordered on each side by a yew hedge [11I] of about twelve feet high. After Sir Henry arrives at the Hall he asks to see this yew alley, which turns out to be *a long, dismal walk*. (The Hound of the Baskervilles).

Yew trees and a hedge also feature in The Valley of Fear, where rows of very ancient yew trees cut into strange designs circle the garden of the Manor House at Birlstone, and at the end farthest from the house they thicken into a continuous hedge. Further trees, this time pollarded [11I] elms lining the village street, and oaks around the driveway of the Manor House are described in the same story. Huge beech trees [11I] also grow in the park surrounding the house. The grounds of Pondicherry Lodge have a corner screened by a young beech in another case (The Sign of Four); and the beech [11I] trees in The Copper Beeches must have given the house in the story its name. Elsewhere (The Sign of Four) Jonathan Small had to work all day under the mangrove trees [11I] when a captive on the Andaman Islands [11H].

One day when Holmes and Watson take a springtime walk in the Park, the first green shoots are appearing on the elms [11I], and the sticky spear-heads on the chestnut trees [11J] are just bursting into leaf. (The Yellow Face). On their return Holmes finds that a visitor has called whilst he and Watson were out, but has left behind a brier [11J] pipe with a long amber [11J] stem.

The Hound of the Baskervilles has numerous references to plants. Stapleton, as a naturalist, carries a tin botanical specimen box [11J], and tells Watson that Grimpen mire is the location where rare plants can be found. Later when Holmes and Watson arrive there they report: *Rank reeds [11J] and lush, slimy water-plants sent an odour of decay and a heavy miasmatic [11J] vapour into our faces.*

Lichens [11J] have blotched the pillars of the lodge-gates of Baskerville Hall, and they are also present at Hatherley Farm, where there are *great yellow blotches of lichen upon the grey walls.* (The Boscombe Valley Mystery).

'Miss' Stapleton asks Watson to fetch an orchid [11J] for her from among the mare's-tails [11J], and remarks that *"We are very rich in orchids on the moor."* White cotton grass [11K] also grows there. Dripping moss [11K] and fleshy hart's tongue ferns [11K] are noticed by Watson as he, Sir Henry Baskerville and Doctor Mortimer travel from the local railway station to Baskerville Hall. Also bronzing bracken [11K] and mottled bramble [11K] are seen along the route. (The Hound of the Baskervilles).

Watson goes into literary mode in another adventure (The Retired Colourman) when describing Josiah Amberley's residence in built-up Lewisham. In *"a little island of ancient culture and comfort, lies this old home, surrounded by a high sun-baked wall mottled with lichens and topped with moss, the sort of wall ..."* But Holmes is not impressed and interrupts him. *"Cut out the poetry, Watson," said Holmes severely. "I note that it was a high brick wall."*

In **The Dying Detective** it is reported that Holmes puts belladonna [11K] in his eyes to give them *the brightness of fever.*

Holmes finds livid [11K] fungi [11K] growing on the inside of a wooden box which is discovered in an underground chamber at Hurlstone.

At another house, Briarbrae, Holmes admires a blooming moss-rose [11K], and looking down at the dainty blend of crimson and green proclaims philosophically: *"Its smell and its colour are an embellishment of life, not a condition of it"*. (The Naval Treaty).

NOTES 11A

Swamp adder: The Speckled Band is a fictitious name, but there is a venomous pit viper (Wagler's) which is speckled with bands of colour. This can be found in swamps in Asia.

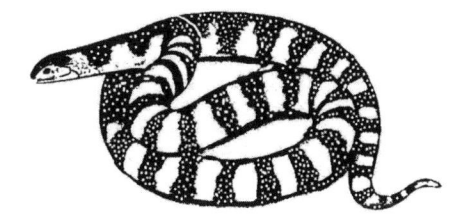

Vipers: Poisonous snakes of the family Viperidae. Some species are better known as adders.

The gila (Gila Monster) is a venomous lizard with a painful bite, though this is seldom fatal. The reptile has a stout scaly, black and pink patterned body and can grow up to about twenty inches long. It can be found in some areas of America and Mexico.

The slow-worm or blindworm is a limbless lizard common in Europe.

Beetles: Many species of winged insects of the order Coleoptera.

The Ganges: India's greatest and longest river which rises in the Himalayas and flows over 1,500 miles to the delta in the Bay of Bengal.

The ichneumon in this story is a **mongoose**, which is a small mammal. The Indian mongoose is grey or greyish-brown and will fight and kill snakes.

The cobra is a large hooded and venomous spitting snake, found in Africa and Asia.

NOTES 11B

Red leech: There are many species of leech, which are classified as Hirudinea within the phylum Annelida (segmented worms). They have suckers to latch on to animals and humans from whom they suck blood in order to obtain sustenance.

Pastiche stories: One notable collection of twelve further stories about Holmes and Watson was published as *The Exploits of Sherlock Holmes* in 1954. The authors were Adrian Conan Doyle (Sir Arthur's son) and John Dickson Carr, a popular writer of crime novels.

The langur is an Indian monkey. It has a slender body and long tail. It is also known as the Entellus Monkey or Hanuman, revered by Hindus.

Anthropoid means 'similar to man'.

Cheetah: A member of the cat family found in Africa and India, and the fastest mammal on earth. It is a good hunter.

Baboon: A ferocious but intelligent animal, which has a snout like a dog, cheek pouches, strong jaws and prominent canine teeth. It is found in Africa (e.g. Ethiopia) and Saudi Arabia, but not in India as implied in the story.

Coyote: It is similar to a jackal, and sometimes known as a Prairie Wolf. It usually lives in burrows.

NOTES 11C

Grizzly bear: A large bear found in the rocky and mountainous regions of the U.S.A. and Canada. 'Grizzly' means grey coloured.

Weasel: A small carnivorous hunter of small mammals such as mice and rabbits.

Stoat: Another name for this small mammal is the Ermine, and this term is used for the fur it provides for the embellishment of some robes and clothes. Animals in colder climes or in the winter can acquire thick white or partially white coats.

Cyanea capillato **(The Lion's Mane)** is a stinging jellyfish which can sometimes be found off the British coast. In the past it was rare but today numbers are increasing slightly. The jellyfish derives its name from its tentacles, which resemble a lion's mane and have been known to reach over 40 feet long on an exceptional specimen.

Taxidermy: The preparation and preservation of dead creatures for display.

Cockroach: A common household pest.

Baron George Léopold Chretién Frédéric Dagobert Cuvier (1769-1832) was a famous French anatomist who devised a comprehensive system of animal classification.

NOTES 11D

Giant rat of Sumatra: Sumatra is an Indonesian island in Asia, and the giant rat may just have been an overgrown specimen of the common brown rat *(Rattus norvegicus)* or the black rat *(Rattus rattus)*. However it has been suggested that an unusual animal called a Moon Rat which can be found on the island, may have been the creature in question.

Mole: A small dark velvet-furred burrowing animal that eats worms and insects.

The bittern: A bird that often favours reed-beds for its habitat. Its very distinctive cry or 'boom' is made only by the male in the breeding season.

The raven: A large black scavenging bird and a member of the crow family. It eats carrion, insects, reptiles and small mammals.

Gulls: There are many species. They are omnivorous and good swimmers and fliers.

Curlew: A wading bird with a long curved bill and a musical cry.

The plover: Probably the Golden Plover in this story. The bird has a fine summer plumage of black and gold.

The buzzard: In this adventure the Turkey Buzzard, an American bird of prey similar to the vulture. It eats carrion, and sometimes plays a waiting game when it senses death is near.

NOTES 11E

Mormons: A religious group also known as the Church of Jesus Christ of Latterday Saints. Founded in the U.S.A. by Joseph Smith in 1830 in New York State. Later, divisions took place in the Church, and in 1847 the main group of Mormons moved to establish a base at Salt Lake City in Utah, under the presidency of Brigham Young.

The woodcock: A woodland bird with a reddish-brown and black plumage. Valued as a game-bird at the time of the adventures.

Game-cock: A specially bred cockerel for cock-fighting.

Voodoo: Superstitions, magic and other rites and beliefs of a religion or cult practised in the West Indies, especially Haiti.

Cock: A cockerel or male bird as compared to a chicken.

Mulatto: A person of mixed race.

Carrion crow: A large bird, black in colour, which feeds on carrion and a wide range of foods.

Carrion: Rotting animal tissue.

The Fulham Road is a long thoroughfare through Chelsea and Fulham in London.

Luminous Paste: When Conan Doyle wrote The Hound of the Baskervilles, a luminous mixture at that time may possibly have contained phosphorus, the luminescence resulting from an oxidative reaction of the phosphorus. Alternatively, a chemical such as impure barium sulphide or calcium sulphide may have been used with the phosphorescence due to the presence of traces of heavy metals in the compound.

NOTES 11F

Spaniel: A popular breed of dog of which there are several varieties.

Terrier: There are many varieties. The Bull Terrier and the Jack Russell are two of the best-known breeds.

Mastiff: A large powerful dog once used for fighting.

Shoscombe breed of spaniels: Fictitious.

Creosote from coal tar was a viscous substance used for treating timbers to protect them from damp rot and pests. Creosote was discovered about 1833 by the German industrialist, Baron Karl von Reichenbach. The name Reichenbach will be familiar to all enthusiasts of Sherlock Holmes stories, as the waterfall site of the life-and-death struggle between the famous detective and his opponent, Professor Moriarty.

Bull pup: A young Bulldog – a sturdy animal often considered a national symbol of Britain showing grit, tenacity and aggression in difficult situations.

Bull terrier: A cross between a bulldog and a terrier.

A draghound is a hound which follows an artificially made trail instead of a fox.

Aniseed comes from the Anise plant, and its powerful and recognizable scent could be followed by dogs.

Wolfhound: This is probably a reference to the Irish Wolfhound, a very large rough-coated dog used in hunting.

NOTES 11G

Airedale terrier: A sporting dog of Yorkshire origin, hence the name.

Cobs: Short-legged, sturdy horses, ideal for riding or driving as long as stamina rather than speed is required.

Wagonette: A four-wheeled open carriage with seats facing each other.

Mustang: A small wild American or South American prairie horse.

Mule: The offspring of a mare and a male donkey.

Guineas: A guinea was 21 shillings (105 pence). Therefore the probable cost of each horse, according to Holmes, would have been one hundred and fifty seven pounds and ten shillings (£157.50).

In Holmes' day the **British Museum** was responsible for the Nation's natural history collections as well as for antiquities.

Butterfly collecting was a fairly common Victorian pastime in an age when wildlife was more abundant. Although there may have been less concern for the environment, there was a keen interest in natural history for its educational value. Killing and mounting the caught specimens was a necessary part of butterfly collecting. Potassium cyanide or chloroform was used for killing the insect. When the insect was dead it had a pin inserted through its thorax, and the butterfly was then fixed onto a wooden or composite wooden/cork setting board, where the wings were arranged outspread for drying and display. A label containing details of the specimen and place and date of capture was an important requirement for scientific study.

NOTES 11H

***Cyclopides*:** This would have been a Skipper butterfly, perhaps the Chequered Skipper which would have been a rare find in this part of England.

Beekeeping and the production of honey have been known for centuries. Early beehives (known as skeps) were made of straw and the structures were dome-shaped. Some wooden Victorian hives were built like tiny houses or cottages with doors and sloping roofs. Modern box-like hives are made of cedar wood. The products from bees are liquid honey, comb-honey and beeswax. The bees use the wax they make in their hives as the structural component of the cell walls of the combs in which the honcy is deposited and the eggs and the young bees develop.

Beeswax has a pleasant smell and is yellowish in colour. It can be used in the preparation of a number of products such as candles, polishes, pharmaceutical preparations, and cosmetics.

Mosquitoes frequently breed in stagnant swampy places, and the bite of the female *Anopheles* mosquito can cause the unpleasant disease of malaria. See also Chapter 2.

The Andaman Islands are situated in the Bay of Bengal.

Agra: A city in the Indian state of Uttar Pradesh.

A scorpion has strong claws and a tail with a curved sting which can cause a poisonous wound.

Centipedes: These are arthropods with long segmented bodies and many pairs of legs. They are nocturnal and carnivorous.

A bibliophile is a book collector.

NOTES 11I

Oaks: The common English oak with its familiar acorns is *Quercus robur*. There are many other varieties. Some grow to a great height and a great age.

Elms: The English elm is *Ulmus procera*, which has tooth-edged leaves and red flowers. The Wych elm was also common, and other species could be found in Britain, but many trees were lost from Dutch elm disease in the 1970's.

Scrub Oak: A stunted oak tree or trees.

Fir: Also known as a Pine tree with its familiar needles and cones.

Yew hedges are found in many gardens and parks and are frequently trimmed and shaped into topiary designs to represent animals, birds and other topics. Yew trees can often be seen in churchyards and may grow to a great height and immense girth, though many are hollow. The bark and needles are poisonous.

Pollarded: The trees have been top cut to allow young branches to grow.

Beech trees: The Common Beech is found throughout Southern England. The Copper Beech gets its name from the coppery red leaves which make it so distinctive.

Mangrove trees grow in swampy tropical places in clumps. The common mangrove has yellow flowers and its pear-shaped fruit is edible.

NOTES 11J

Chestnut trees: The best known are the Horse Chestnut with its conkers in prickly 'cases', and the Sweet or Spanish Chestnut which yields edible nuts for roasting, marron glacés or flour-making.

Brier (or briar) pipes are made from the root of the Tree Heath *Erica arborea*, which is grown in the Mediterranean region of Europe and other parts of the world. The wood is suitable for pipe making as it is very hard, has a fine texture which can be highly polished and chars only slowly.

Amber is a hard fossilized resin the colour of golden honey and contains succinic acid. The remains of ancient creatures or plants are sometimes found preserved in amber.

Botanical specimen box: Plants are placed in this container (known as a vasculum) when collecting in the field in order to preserve them and prevent damage.

Reeds: Tall waterside or marshland growing plants. Some species can be used for thatching.

Miasmatic: Unpleasant or poisonous aroma from decaying vegetation.

Lichens are unusual plants. Each one consists of two species (algae and fungi) living symbiotically together – that is, to the mutual benefit of each other. Lichens can endure heat and cold easily and thrive on many surfaces such as tree bark, rocks and masonry, as well as on the ground.

Orchid: There are both common and rare varieties of this large family of plants, the *Orchidaceae*. Several species have lipped or spectacular flowers which are bright and colourful.

Mare's-tails: An aquatic herb which has medicinal uses.

NOTES 11K

Moss: There are numerous varieties. The plants have a cellular structure and are non-flowering. They reproduce by means of spores, and can grow on almost any surface including brickwork, tiles, and rock.

Hart's tongue ferns: A widely distributed species with tufts of long tongue-shaped fronds. Ferns are flowerless, cellular plants with reproductive spores.

Bracken: A common fern with large triangular-shaped fronds.

Bramble: The ubiquitous blackberry plant. Bramble jelly can be made from the berries.

White cotton grass: Found on boggy ground, especially moorland. The plants have white filaments similar to tufts of cotton.

Belladonna is another name for the Deadly Nightshade plant, which is extremely poisonous, especially if the berries are consumed. However it is also a very useful plant, as from it can be obtained the alkaloid drug belladonna, which has various medical uses. In addition the important drugs atropine and hyoscine (scopolamine in the U.S.A.) are derived from this plant.

'Livid' is a black/blue colour.

Fungi: Unlike many plants, fungi lack chlorophyll and are unable to photosynthesize. To some extent they are parasites or saprophytes and depend on either living or dead material (the rotten wooden box lid in this case) to flourish.

The moss-rose is so named because the calyx has a moss-like growth. The flower is very fragrant and there are several varieties.

The Blue Carbuncle

12 "YOU KNOW MY METHODS WATSON!"

DEDUCTION AND FORENSIC SCIENCE

In the words of Sherlock Holmes:

Like all other arts, the Science of Deduction and Analysis is one which can only be acquired by long and patient study, nor is life long enough to allow any mortal to attain the highest possible perfection in it. (A Study in Scarlet)

Reference has already been made in the introduction to the two methods of scientific deduction that Holmes employs in his investigations. Briefly these are the use of known facts and collected data, and secondly, the subjection of samples to scientific tests or microscopic examination in the laboratory. However as both methods can frequently overlap, they are being treated in combination in some parts of this chapter.

The Book of Life is a journal article (mentioned in A Study in Scarlet) written by Holmes, that attempts *to show how much an observant man might learn by an accurate and systematic examination of all that came his way.* Scrutinising people closely can be a good exercise he suggests. *It sharpens the faculties of observation, and teaches one where to look and what to lo look for.*

Holmes examined, one by one, the articles which Lestrade had handed to him (The Cardboard Box).

Numerous illustrations of Sherlock Holmes depict him wearing a deerstalker [12A] and holding a magnifying glass. Comment here on his choice of headgear is unnecessary. Probably the deerstalker was not the usual type of hat that the fashionable detective of the time would always wear, and indeed Holmes only chose to wear a hat of this type when travelling or in the country. However the importance of the magnifying glass cannot be overlooked, as it was an essential aid when Holmes investigated a case.

Holmes uses his magnifier to find relevant clues to support his theories and ideas about the cases that he attends. Although some clues, such as wheeled tracks and footprints, are large enough to see clearly without magnification, the magnifier is a great aid to enlarging the size of patterns and ridges created by vehicles and shoes, or for finding small objects dropped by an assailant or victim. Small drops of blood, types of soil and marks of all varieties can also be scrutinized with its help. For example Holmes uses his magnifying glass to find dust marks on the stair carpet at the home of Bartholomew Sholto (The Sign of Four). (Some further examples are given in Chapter 9).

No police force could operate efficiently today without a permanent visual image of all important tangible pieces of evidence, crime scenes, injuries, bodies of victims in situ, and so on. An example of acquiring forensic evidence using a camera to effect by Holmes occurs in the case of The Lion's Mane (see Chapter 9).

CARRIAGE WHEELTRACKS

In his regular scrutiny of such tracks, Holmes is aided by the frequently soft, muddy and probably mucky (from horse droppings) condition of the road surfaces of the period. The tracks of a cab in A Study in Scarlet are noted by Holmes, who declares that: *"The very first thing which I observed on arriving there was that a cab had made two ruts with its wheels close to the curb."* At a later point in the same adventure, Holmes again comments on the cab, telling Watson that *"I satisfied myself that it was a cab and not a private carriage by the narrow gauge of the wheels. The ordinary London growler*[12A] *is considerably less wide than a gentleman's brougham*[12A]*."*

Wheel tracks also appear in The Greek Interpreter, and Holmes makes the following inference from his observation of these: *"A carriage heavily loaded with luggage has passed out during the last hour."* He has seen that the outward bound tracks are deeper than the inward bound ones, hence making the logical deduction that the departing coach was heavily laden.

FOOTPRINTS AND OTHER PRINTS

These appear in several places in the canon. In one story (A Study in Scarlet) Holmes remarks on the observation of human tracks: *"There is no branch of detective science which is so important and so much neglected as the art of tracing footsteps. Happily, I have always laid great stress upon it, and much practice has made it second nature to me."* Holmes then states that two men have passed through the garden, one tall because of the length of his stride, and the other well-dressed because of his small elegant boot impressions.

In another of the adventures (The Boscombe Valley Mystery), Holmes takes great pains to measure the boots of Charles McCarthy and his son very carefully from seven or eight different points. At Boscombe pool he carefully examines any tracks that remain, in order to work out the sequence of events leading up to the murder of McCarthy senior, and the details of what occurred afterwards. *"These are young McCarthy's feet. Twice he was walking, and once he ran swiftly so that the soles are deeply marked, and the heels hardly visible."*

The many analyses of footprints [12A] in the stories follow a similar pattern of close scrutiny and then deduction. They appear in The Crooked Man along with some workman's boot-nail marks which Holmes notices in Watson's own home; while in The Sign of Four Holmes actually reveals that he has written a monograph *'Upon the tracing of footprints, with some remarks upon the uses of plaster of Paris* [12A] *as a preserver of impresses'*.

In The Reigate Puzzle there is an interesting variation from the usual observation, in that Holmes finds a *lack* of boot marks in a place where there should certainly have been some, if the testimony of the Cunninghams about an escaping man can be believed.

A handprint is found in The Sign of Four as Holmes clambers up on to a wall and discovers a print of Wooden-leg's hand. Out of context this phrase reads strangely, but 'Wooden-leg' is a soubriquet for Jonathan Small. In the same adventure Holmes tells Watson that he can often detect the trade of an individual from the form of his hand, and then goes on to show Watson a monograph *with lithotypes* [12A] *of the hands of slaters, sailors, cork-cutters, compositors, weavers, and diamond-polishers,* that he has published upon the subject. Macabrely he points out that such information is of great value in the identification of unclaimed bodies. However such methods today would be highly suspect. What would Holmes have made of the hands of workers in present day occupations, such as supermarket check-out operators, call centre staff and financial advisers to mention just a few examples?

A thumb-print is found in The Sign of Four, and a thumb-print or thumb-mark in The Norwood Builder. The latter case, where John Hector McFarlane is suspected of murder, has Holmes actually matching a blood stained print on a wall with a *"wax impression of young McFarlane's right thumb."* Before this though, he says to Inspector Lestrade: *"You are aware that no two thumb-marks are alike"* [12A], and Lestrade replies: *"I have heard something of the kind."* The matching thumb-print however is all part of an ingenious false trail, as a reading of the story will reveal.

ANIMAL TRACKS

Footprints of animals, whenever they are found in the course of an investigation and deemed to be important, receive the same treatment of observation and deductive reasoning. In The Crooked Man Holmes shows Watson a piece of tissue paper. *"What do you make of that?"* he asked. *The paper was covered with the tracings of the footmarks of some small animal. It had five well-marked footpads, an indication of long nails, and the whole print might be nearly as large as a dessert-spoon. "It's a dog," said I.* Holmes airily dismisses this answer as he has found traces to indicate that the animal had climbed a curtain, which he says a dog would be unable to do. He then proceeds to give a detailed account of the measurements of the creature and deduces that *it was probably some creature of the weasel* [12B] *or stoat* [12B] *tribe.* Near the end of the story we learn that the animal was in fact a mongoose [12B].

Silver Blaze is a race horse that disappears from its stables, and when Holmes sets out on the investigation, he finds a hoof print that he compares to a lost shoe from the animal. *The track of a horse was plainly outlined in the soft earth in front of him, and the shoe which he took from his pocket exactly fitted the impression.* Both he and Watson continue to follow the missing horse's trail until it takes them to the Capleton (or Mapleton in some editions) stables of Silas Brown. (Silver Blaze)

A horse that draws a cab helpfully leaves its hoofprints in the wet ground for Holmes to spot in A Study in Scarlet. *There were the marks of the horse's hoofs, too, the outline of one of which was far more clearly cut than that of the other three, showing that that was a new shoe.*

The footprints of the 'ghostly' hound are like no others from a supposed apparition, meaning that they can be seen, and indeed they are – by Doctor Mortimer. After Sir Charles Baskerville's death, the doctor observes some footprints near the body. When Holmes later asks him whether it was: *"A man's or a woman's?"* Watson records the dramatic reply. *Dr. Mortimer looked strangely at us for an instant, and his voice sank almost to a whisper as he answered: "Mr. Holmes, they were the footprints of a gigantic hound!"* (The Hound of the Baskervilles).

DEDUCTIVE OBSERVATIONS

The height of suspects and witnesses is an important element in the description of any person sought in connection with a crime. One method that Holmes uses, is to work out the height from the length of the person's stride. He does this in A Study in Scarlet, and peremptorily tells Watson that *"It is a simple calculation enough, though there is no use my boring you with figures"*.

Again in The Boscombe Valley Mystery Watson refers to this method of assessing height, but Holmes goes further in his inferences and interprets lameness from a man's footprints. Watson asks how he reaches this conclusion and Holmes explains. *"The impression of his right foot was always less distinct than his left. He put less weight upon it. Why? Because he limped – he was lame."* In addition, from the way the blow was struck the murderer is identified by Holmes as being left-handed.

In another case (The Sign of Four) Holmes deduces Jonathan Small's height in the same way; and earlier he impresses the impulsive Scotland Yard detective Athelney Jones, with his detailed, and as it turns out, accurate description of Small, composed from the various clues that he has observed. It is interesting to find that Detective Jones sneers at Holmes' methods, giving him the nickname of 'Mr. Theorist'. Conan Doyle, as a good writer, created friction between the characters, and the discernible tension between Holmes and the official detectives at times is expertly supplied by him.

Holmes makes a number of deductions from a hat in The Blue Carbuncle, but first of all he hands over his lens to Watson to see what he can make of it: *"What can you gather from this old battered felt[12B]?"* Poor Watson looks at it carefully but has to answer: *"I can see nothing"*. Holmes then confounds him with an extremely long list of the characteristics of the owner of the hat. The skill of the detective is impressive, but we should not forget that it was Conan Doyle who devised this astonishing series of clues for Holmes to evaluate.

Time and again Holmes likes to show off his deductive skills. In The Boscombe Valley Adventure he surprises Watson by telling him that he knows that *"in your bedroom the window is upon the right-hand side"*. He has inferred this from the way in which Watson shaved that morning, noting that one side of Watson's face was smoother than the other, meaning that the quality of the shave was comparable to the amount of light falling on that side of the face. In another story (A Case of Identity), he reproves Watson for missing important points about Miss Sutherland: *"I can never bring you to realize the importance of sleeves, the suggestiveness of thumb-nails, or the great issues that may hang from a bootlace."* A further adventure (The Red-Headed League) provides another good example of Holmes' deductive powers. He sees that Jabez Wilson has *"at some time done manual labour, that he takes snuff*[12B]*, that he is a Freemason*[12B]*, that he has been in China, and that he has done a considerable amount of writing lately"*. This is of course all explained in the story, but the link with China is interesting. Holmes tells Wilson that his own contributory knowledge of tattoo marks has been published, and that the tattoo[12B] of a fish on Wilson's wrist could only be of Chinese origin. *"That trick of staining the fishes' scales of a delicate pink is quite peculiar to China."* He also says that a Chinese coin on Wilson's watch-chain confirms his point.

Another tattoo, or more accurately brand-mark, is mentioned in The Valley of Fear and is observed to be a triangle inside a circle on Mr. Douglas's forearm. The detective is intrigued by this. On another occasion Holmes' examination of a room where a crime has taken place and the lawn outside the house (The Crooked Man), provides him with the clues of footmarks of both a person and an animal. *"There had been a man in the room, and he had crossed the lawn coming from the road. But it was not the man who surprised me. It was his companion."*

The Yellow Face presents Holmes with a difficult situation. After investigation he makes assumptions, but at the end has to admit that they were wrong. As a result he asks Watson to remind him of this case *"if it should ever strike you that I am getting a little over-confident in my powers."* What a relief, Holmes has some humility after all!

FORENSIC DEDUCTION

This type of deduction needs more than just simple observation, it can require analysis by means of chemical tests, dissection, and/or microscopic study. (See also Chapter 6)

On one occasion (Silver Blaze) Holmes is invited to investigate a 'horsey' case, some years before Dick Francis's heroes and investigators come onto the scene. An analysis of the stable-lad's supper is carried out after Hunter, the lad in question, is found drugged. The remains of the meal are found to contain powdered opium [12B], and in the same story, a wax vesta [12B] is discovered in the mud by Holmes. He says that he was looking for it, a remark which surprises Inspector Gregory, who is working on the case for the police. Perhaps the most interesting deduction in this adventure, is the one that Holmes makes from something that did **not** happen. *"Obviously the midnight visitor was someone whom the dog knew well,"* because the dog did not bark, explains Holmes after solving the case.

By close examination, Holmes finds scratches on the ward [12B] of the door lock of the room in which Mr. Blessington is found hanged. The murderers *"with the help of a wire, forced round the key,"* he states. (The Resident Patient). Signs of forced entry are also found in another adventure (The Reigate Puzzle). *It was evident that a chisel or strong knife had been thrust in, and the lock forced back with it. We could see the marks in the wood where it had been pushed in.*

While evidence of 'breaking and entering' can be fairly easily detected, smells are not always suspected as being of importance. However they provide the breakthrough clues in The Retired Colourman [12B]. On solving this case, Holmes points out that Watson observed the smell of paint but failed to draw the inference, and also that a police officer had smelt gas, the medium used to murder the victims. Holmes being more sensitive to his surroundings, explains that the paint was used to mask the strong smell of the gas [12C].

While Holmes takes two hours to examine minutely the dining-room at

Abbey Grange, Watson sits and watches, knowing that Holmes will be able to deduce a great deal from the clues that he finds. He is astonished though to see the detective climb onto the mantelpiece, albeit to take a closer look at the few inches of red cord which were still attached to the wire. The red cord is a bell-rope torn down by the murderer and used to tie Lady Brackenstall to a chair. The wire is part of the bell-pull system for summoning servants. By close inspection of the end of the cord which he cannot reach, Holmes explains to Watson how he knows the size of the man for whom they are looking: *"I could not reach the place by at least three inches – from which I infer that he is at least three inches a bigger man than I."* Further deductions follow as Holmes pieces together exactly what has happened at the Abbey Grange (the title of the story).

The Valley of Fear contains an interesting clue. At the scene of the crime, the length of a consumed candle [12C] is assessed to indicate the approximate amount of time that passed before the attack took place. Of course it was essential for the candle to be a new one, so that the amount consumed could fairly accurately be measured. Otherwise its exact length would have had to have been known before it was lit – clearly a most unlikely situation.

"When you have eliminated the impossible, whatever remains, however improbable, must be the truth." (The Sign of Four) With these words Holmes demonstrates to Watson that the killer/s in this mystery must have come through a hole in the roof, as all the other ways of entry or concealment have been eliminated as being impossible. There are undoubtedly two assailants and both have left traces which yield clues that enable Holmes to list comprehensive descriptions of them.

The remains of burning are important in two cases [12C]. In The Devil's Foot Holmes puts ashes which he has scraped from the lamp into an envelope for analysis; whereas at Lower Norwood [12C], charred organic remains are found after a fire at the premises of Mr. Oldacre, together with several discoloured metal discs discovered in the ashes.
(The Norwood Builder) (see also Chapter 6).

BLOODSTAINS

It goes without saying that blood and bloodstains in criminal cases are a common indication of injury and foul play. They naturally appear in many of the stories especially when a murder has taken place.

In A Study in Scarlet there is a message on the wall written in blood: a dramatic gesture by a murderer intent on misleading the police. Tobias Gregson, a Scotland Yard detective, has written to Holmes requesting his assistance in this case of baffling clues, and the great detective is willing to oblige.

In two of the stories knives [12C] are found with blood on them. In The Six Napoleons Beppo is arrested for breaking and entering a house, and when searched, a long sheath-knife [12C] is found on him, *the handle of which bore copious traces of recent blood.*

A bloodmark is discovered in The Sign of Four; and blood is found on the yellow gorse blooms in The Priory School, as well as on the path and heather growing close by. In Black Peter Holmes observes a discolouration on a notebook found at the scene of the crime. Inspector Hopkins confirms that it is a bloodstain, and Holmes establishes the fact that it was dropped after the crime as the stain is on the underside of the book.

In the case of The Norwood Builder, a newspaper account of the crime at Mr. Oldacre's house, reports *that there were signs of a murderous struggle, slight traces of blood being found within the room.* When Holmes goes there to investigate he finds that the stains are *"very slight, mere smears and discolourations, but undoubtedly fresh."* Bloodstains also appear in various other stories [12C]. They are found on some carpet slippers belonging to Mr. Barker, a friend of Jack Douglas, whose death Holmes is investigating in Sussex. Holmes matches one of the slippers with a bloodstain on the window sill, the whole procedure exciting the accompanying local police officer, White Mason, who, in referring to the case, proclaims: *"I said it was a snorter!"* [12C] (The Valley of Fear).

Blood shows up very well on snow, and this is where Holmes finds a few drops in The Beryl Coronet, enabling him to follow a trail as far as the main road, which had unfortunately been cleared, *"so there was an end to that clue,"* he reports. In the case of The Second Stain, the bloodstain is perhaps the most important clue, because the discolouration on the carpet in the room where the murder has taken place, does not correspond with the stain on the wooden floor beneath the carpet. Holmes finds that the carpet has been turned round since the crime, and he has to find out why and by whom.

The Sign of Four has a scene where Holmes finds a stout rope still in situ at the scene of the crime. He observes through his lens some bloodmarks near the bottom of the rope, and concludes that the climber, when descending, *"slipped down with such velocity that he took the skin off his hands."*

The Lion's Mane case has an entirely different scenario in that the origin of the bloodstains on the victim are not inflicted by anyone with criminal intent. Holmes is puzzled by the death of Fitzroy McPherson, and discusses with Inspector Bardle the appearance of some weals on the deceased's body. *"There is a dot of extravasated* [12C] *blood here, and another there. There are similar indications in this other weal down here. What can that mean?"* True to form, as in so many investigations, Holmes soon finds the answer and satisfactorily solves the mystery.

As the foregoing pages have shown, Holmes' methods of deductions were based on data and observation combined with forensic knowledge and the ability to carry out practical tests under laboratory conditions. As many of his cases were solved using these techniques, we can conclude that his methods of detection were highly effective, and usually enough evidence could be found for even the cleverest of criminals to be caught and brought to justice.

NOTES 12A

Deerstalker: A double-peaked cloth cap, with ear flaps tied on top.

The Growler or Clarence cab was a four-wheeled horse-drawn vehicle in use at the time. The popular design of this cab dates from the 1830's (though some four-wheeled cabs were in use in the 1820's), and required one or two horses to pull the vehicle. It usually carried four passengers. Growler was a London slang name for this cab.

The Brougham was a private four-wheeled carriage named after Lord Brougham who introduced it in 1838. The advantage of it was its lightness so that it only required one horse. There were different models and some were given nicknames. The 'pill box' was said to be the doctor's favourite!

Plaster of Paris is a partially hydrated form of calcium sulphate. It sets very rapidly when water is added and therefore is very convenient for producing casts and mouldings. (See also Notes 6F in Chapter 6).

Lithotypes: The illustrations in Holmes' publications would probably have been reproduced from etched stone printing plates.

Footprints: Stories with reference to footprints include: The Dancing Men; The Norwood Builder; The Beryl Coronet; Silver Blaze; The Resident Patient; A Study in Scarlet; The Valley of Fear; The Naval Treaty.

Fingerprints: The importance of fingerprints for identity purposes was recognized by Sir William James Herschel, a civil servant in India in 1858. The first police force to make use of fingerprints for criminal investigation was a division of the Argentinian Buenos Aires force in 1891. The officer who introduced the system was Juan Vucetich, and he soon proved the advantages of the classification scheme devised by him and utilized there. Two Britons were also instrumental in the development of the science of fingerprints. Sir Francis Galton studied the identification of prints and published three works on them; and Sir Edward Henry, who became Commissioner of the Metropolitan Police, invented his own system of classifying prints, which was instigated in India in 1897 and a few years later in London under the control of a Central Fingerprint Bureau. All these schemes of course relied on the almost complete uniqueness of every print that was to be recorded or compared.

NOTES 12B

Weasel: a small carnivorous hunter of small mammals such as mice.

Stoat: Another name for this small mammal is the ermine, and this term is used for the fur it provides for the embellishment of some robes and clothes. Animals in colder climes or in the winter can acquire thick white or partially white coats.

Mongoose: The Indian mongoose is a grey or greyish-brown small mammal which will fight and kill snakes.

'Felt' refers to a felt hat made of cloth consisting of compressed wool.

Snuff: information about snuff can be found in Chapter 4.

Freemasons: A fraternal secret society based on area Lodges, with the aim of mutual benefit to members in both business and everyday life as exemplified by the maintenance of institutions for the education and health needs of members' families. Features of the Freemasons are progressive orders of seniority, initiation rites and secret signs of recognition between members.

Tattooing has a long history and became popular in Britain in the 19th century. A tattoo can be an artistic design, pattern or identification mark on the skin in black or coloured ink or dye (though carbon and gunpowder in water have also been used). A suitable instrument of incision, usually a needle, is required to carry out the work.

Powdered opium: opium is prepared in powder form from the opium poppy. Morphine is derived from it and used in medicine as a pain killer. Opium was popularized in Britain from about 1680 by the physician Thomas Sydenham. It became available in laudanum in Victorian times.

Wax vestas were matches which could be struck anywhere. They were cotton tapers or wooden splints coated in wax to make them burn easily. The heads of early versions consisted mainly of phosphorus, but later ones usually contained phosphorus sesquisulphide and potassium chlorate. Substances such as ground glass, an adhesive and a metal oxide were also often added.

Ward of a lock: The internal notches and fixed metal parts allowing an individual key to unlock the lock.

A colourman: A dealer in paints and brushes.

NOTES 12C

Coal gas: In those days coal gas, which had a stronger smell than today's natural gas and, being mainly carbon monoxide, was lethal if inhaled in high volume in a closed atmosphere.

Candles: Before clocks were widely available, candles were often marked with gradations along their length, each mark representing the passage of one hour.

Cases where **burnt remains** are important: The Devil's Foot; The Norwood Builder.

Lower Norwood is a suburb of South London.

Knives: Stories with references to knives: The Six Napoleons; Silver Blaze.

Sheath-knife: A sharp bladed knife carried in a case or cover often made of leather.

Bloodstains: Stories with references to bloodstains include: The Beryl Coronet; The Second Stain.

Snorter: A slang term for a difficult, extraordinary or intriguing situation.

Extravasated: A medical term meaning that blood has leaked into close surrounding tissue.

LIST OF SHERLOCK HOLMES STORIES

The series of Sherlock Holmes stories (known as the canon) comprise 56 short stories and 4 novels. The year of first publication of each story, using only its short title and omitting 'The Adventure of ...' is given below. These titles (with some variants) are those usually found in modern collected editions of the adventures.

TITLE	DATE
A Study in Scarlet (novel)	1887
The Sign of Four (novel)	1890
The Boscombe Valley Mystery	1891
A Case of Identity	1891
The Five Orange Pips	1891
The Man with the Twisted Lip	1891
The Red-Headed League	1891
A Scandal in Bohemia	1891
The Beryl Coronet	1892
The Blue Carbuncle	1892
The Copper Beeches	1892
The Engineer's Thumb	1892
The Noble Bachelor	1892
Silver Blaze	1892
The Speckled Band	1892
The Cardboard Box	1893
The Crooked Man	1893
The Final Problem	1893
The "Gloria Scott"	1893
The Greek Interpreter	1893
The Musgrave Ritual	1893
The Naval Treaty	1893
The Reigate Puzzle or The Reigate Squires	1893
The Resident Patient	1893
The Stockbroker's Clerk	1893
The Yellow Face	1893

TITLE	DATE
The Hound of the Baskervilles (novel)	1901-02
The Blanched Soldier	1903
The Dancing Men	1903
The Empty House	1903
The Norwood Builder	1903
The Solitary Cyclist	1903
Black Peter	1904
Charles Augustus Milverton	1904
The Priory School	1904
The Abbey Grange	1904
The Golden Pince-Nez	1904
The Missing Three-Quarter	1904
The Second Stain	1904
The Three Students	1904
Wisteria Lodge	1908
The Bruce-Partington Plans	1908
The Devil's Foot	1910
The Disappearance of Lady Frances Carfax	1911
The Adventure of the Red Circle	1911
The Adventure of the Dying Detective	1913
The Valley of Fear (novel)	1914-15
His Last Bow	1917
The Mazarin Stone	1921
The Problem of Thor Bridge <u>or</u> Thor Bridge	1922
The Creeping Man	1923
The Illustrious Client	1924
The Sussex Vampire	1924
The Three Garridebs	1924
The Lion's Mane	1926
The Retired Colourman	1926
The Three Gables	1926
Shoscombe Old Place	1927
The Veiled Lodger	1927

Illustrations: Source and Acknowledgements

Cover: (Valerie Naggs), (Sidney Paget)
  Map reproduced by permission of Geographers A-Z Map Co. Ltd.,
  © Crown Copyright 2006. All Rights reserved. Licence No. 100017302
Arthur Conan Doyle, Page 10 (The Harry Price Library, London).
Chapter One: Page 14, 15, 17, 21, 24 (Sidney Paget); Page 16 (Valerie Naggs); Page 26 (Richard Guttschmidt).
Chapter Two: Page 30, 49 (Sidney Paget); Page 40 (Richard Guttschmidt).
Chapter Three: Page 51 (Sidney Paget).
Chapter Four: Page 65 (Fredric Dorr Steele); Page 68 (Josef Friedrich); Page 70 (Frank Wiles); Page 79 (Sidney Paget).
Chapter Five: Page 80 (Sidney Paget); Page 82, 86, 87 (Josef Friedrich)
Chapter Six: Page 97, 103, 111 (Richard Guttschmidt); Page 99 (Sidney Paget).
Chapter Seven: Page 112, 119 (Frank Wiles); Page 123 (Sidney Paget).
Chapter Nine: Page 132, 135 (Frank Wiles); Page 134 (Sidney Paget); Page 139 (Josef Friedrich).
Chapter Ten: Page 140, 141, 142, 144 (Sidney Paget); Page 151 ((Josef Friedrich); Page 160 (Myrtle Bloomfield).
Chapter Eleven: Page 161 (Josef Friedrich); Page 166, 168, 169, 171; Page 172, 186 (Sidney Paget).
Chapter Twelve: Page 187 (Richard Guttschmidt): Page 192 (Sidney Paget).

All other illustrations are provided by the Hadley Pager Picture Library

The publisher has endeavoured to contact all copyright holders and apologises for any inadvertent errors or omissions, and would be pleased to acknowledge these in future editions.

HADLEY PAGER INFO
PUBLICATIONS

Hadley Pager Info has been publishing books since 1996, most of which are French-English and English-French Dictionaries, Glossaries and Phrasebooks of value to British visitors and residents in France, and covering subjects as diverse as Motoring, Gardening and Horticulture, Legal, Medical and Health, Building and Renovation, Veterinary Terms and Conversation in French.

Details of these and our other forthcoming books can be found on our website **www.hadleypager.com**

Our books can be ordered through good bookshops in the UK, as well as through many internet bookshops, or directly from us by sending the details together with payment to Hadley Pager Info, PO Box 249, Leatherhead, KT23 3WX, England. (Postage is free within the UK but a charge is made for delivery outside the UK, see our website for details.)

Our current booklist is available on request through the above PO Box Number or by e-mail to hpinfo@aol.com.